SpringerBriefs in Electrical
and Computer Engineering

SpringerBriefs in Speech Technology

For further volumes:
http://www.springer.com/series/10059

Sreenivasa Rao Krothapalli
Shashidhar G. Koolagudi

Emotion Recognition using Speech Features

 Springer

Sreenivasa Rao Krothapalli
School of Information Technology
Indian Institute of Technology
Kharagpur, West Bengal
India

Shashidhar G. Koolagudi
Department of Computer Science
Graphic Era University
Dehradun, Uttarakhand
India

ISSN 2191-8112   ISSN 2191-8120 (electronic)
ISBN 978-1-4614-5142-6   ISBN 978-1-4614-5143-3 (eBook)
DOI 10.1007/978-1-4614-5143-3
Springer New York Heidelberg Dordrecht London

Library of Congress Control Number: 2012951407

Printed on acid-free paper

Springer is part of Springer Science+Business Media (www.springer.com)

Preface

During production of speech human beings impose emotional cues on the sequence of sound units to convey the intended message. Speech without emotional information is unnatural and monotonous. Most of the existing speech systems are able to process studio recorded neutral speech. However, in the present real world communication scenario, speech systems should have the ability to process the embedded emotions. Emotional clues present in the speech may be observed in various features extracted from excitation source, vocal tract system and prosodic components of speech.

This book attempts to discuss the methods to capture the emotion specific knowledge through excitation source, vocal tract and prosodic features extracted from speech. Various emotion recognition models are developed using auto-associative neural networks, support vector machines and Gaussian mixture models. Emotional speech database in an Indian language Telugu IITKGP-SESC (Indian Institute of Technology Kharagpur-Simulated Emotion Speech Corpus) and Berlin emotional speech database Emo-DB are used in this study for evaluating the emotion recognition performance.

This book is mainly intended for researchers working on emotion recognition from speech. This book is also useful for the young researchers, who want to pursue the research in speech processing using basic excitation source, vocal tract and prosodic features. Hence, this may be recommended as the text or reference book for the postgraduate level advanced speech processing course. The book has been organized as follows:

Chapter 1 introduces speech emotion recognition as an important research area. Psychological and engineering aspects of emotions and their manifestations in speech have been discussed. Chapter 2 provides the review of the literature on speech emotion recognition in view of various emotion-specific speech features, databases and models. This chapter also specify the scope and limitations of the present work. Chapter 3 presents various excitation source features for discriminating the emotions. Chapter 4, briefly discuss about the emotion discrimination capability of vocal tract system features. Chapter 5 discuss about the emotion-specific characteristics present in global and local prosodic features. Chapter 6 summarizes the contents of the book, highlights the contributions of the chapters and discusses the scope for future work.

Many people have helped us during the course of preparation of this book. We would especially like to thank all professors of School of Information and Technology, IIT Kharagpur for their moral encouragement and technical discussions during course of editing and organization of the book. Special thanks to our colleagues at, Indian Institute of Technology, Kharagpur, India and Graphic Era University, Dehradun, India for their cooperation and coordination to carry out the work. We are grateful to our parents and family members for their constant support and encouragement. Finally, we thank all our friends and well-wishers.

Kharagpur, India
Dehradun, India

Sreenivasa Rao Krothapalli
Shashidhar G. Koolagudi

Contents

Acronyms

AANN	Autoassociative Neural Network
CART	Classification and Regression Tree
CV	Consonant Vowel
DCT	Discrete Cosine Transform
DFT	Discrete Fourier Transform
FD-PSOLA	Frequency Domain Pitch Synchronous Overlap and Add
F	Female
FFNN	Feedforward Neural Network
GC	Glottal Closure
Hi	Hindi
HNM	Harmonic plus Noise Model
IDFT	Inverse Discrete Fourier Transform
IITM	Indian Institute of Technology Madras
IPCG	Inter Perceptual Center Group
IPO	Institute of Perception Research
ITRANS	Common transliteration code for Indian languages
LP	Linear Prediction
LPCs	Linear Prediction Coefficients
LP-PSOLA	Linear Prediction Pitch Synchronous Overlap and Add
M	Male
MOS	Mean Opinion Score
NN	Neural Network
OLA	Overlap and Add
PF	Phrase Final
PI	Phrase Initial
PM	Phrase Middle
PSOLA	Pitch Synchronous Overlap and Add
RFC	*rise/fall/connection*
RNN	Recurrent Neural Network
SOP	Sums-of-Product

STRAIGHT	Speech Transformation and Representation using Adaptive Interpolation of weiGHTed spectrum
SVM	Support Vector Machine
Ta	Tamil
TD-PSOLA	Time Domain Pitch Synchronous Overlap and Add
Te	Telugu
TTS	Text-to-Speech
VOP	Vowel Onset Point
V/UV/S	Voiced/Unvoiced/Silenc
WF	Word Final
WI	Word Initial
WM	Word Middle

Chapter 1
Introduction

Abstract This chapter briefly discuss about the need for processing of emotions from speech signal. The chapter begins with the discussion on significance of emotions from psychological and engineering view points. Influence of emotions on the characteristics of speech production system is briefly mentioned. Role of various features extracted from excitation source, vocal tract system and prosody, arc discussed in the context of developing various speech systems. Different types of emotional speech databases used for carrying out various emotion-specific tasks are briefly discussed. Various applications related to speech emotion recognition are mentioned. Important state-of-the-art issues prevailing in the area of emotional speech processing are discussed at the end of the chapter along with a note on the organization of the book.

Speech is a complex signal containing information about message, speaker, language, emotion and so on. Most of existing speech systems such as speech recognition, speaker recognition and speech synthesis, process studio recorded read speech effectively: however, their performance is poor in the case of natural speech. This is due to the difficulty in modeling and characterizing emotions present in natural speech. Presence of emotions makes the speech more realistic and natural. Humans extensively use emotions to express their intentions through speech [1]. The same textual message would be conveyed with different semantics (meaning) by incorporating appropriate emotions. Humans understand the intended message by perceiving the underlying emotions in addition to phonetic information. Therefore, there is a need to develop speech systems that can process emotions along with the message [2]. The basic goals of emotional speech processing are (a) recognizing the emotions present in speech and (b) synthesizing the desired emotions in speech according to the intended message. From the machine's perspective, understanding speech emotions can be viewed as the classification or discrimination of emotions. Synthesis of emotions can be viewed as incorporating the emotion-specific knowledge during speech synthesis. The emotion-specific knowledge is acquired from emotion models, designed for capturing the emotion-specific characteristics.

S.R. Krothapalli and S.G. Koolagudi, *Emotion Recognition using Speech Features*,
SpringerBriefs in Electrical and Computer Engineering,
DOI 10.1007/978-1-4614-5143-3_1, © Springer Science+Business Media New York 2013

1.1 *Emotion*: Psychological Perspective

Emotions have been studied in several scientific disciplines, such as: Biology (Physiological), Psychology, Speech science, Neuroscience, Psychiatry, Anthropology,
Sociology, Communication and so on. Subjects like Business Management and
Advertising need extensive use of emotion processing. As a result, distinctive
perspectives on the concept of *emotion* have emerged, appropriate to the complexity
and variety of the emotions. However, it is important to consider these different
perspectives not as competitive but as complementary. In this book, emotions are
analyzed in view of psychological, physiological and speech perspectives.

Psychology of emotions can be viewed as a complex experience of consciousness
(psychology), bodily sensation (physiology), and behavior (action-speech). The
emotions generally represent a synthesis of subjective experience, expressive
behavior, and neurochemical activity. There are more than 300 crisply identified
emotions by researchers [3, 4]. However generally, all of them are not experienced
in day-to-day life. In this regard, most researchers agree on the principles of *Palette
theory* that quotes, *any emotion is the composition of six primary emotions as any
color is the combination of 3 primary colors* [5]. Anger, disgust, fear, happiness,
sadness and surprise are considered as the primary or basic emotions by most
researchers [6]. These are also referred as *archetypal emotions* [7].

In psychology, expression of emotions is considered as a response to stimuli
that involves characteristic physiological changes [8, 9]. According to physiology,
an emotion is defined as a disruption in the homoeostatic baseline [8]. Based on
these changes, the properties of emotions can be explained in a three-dimensional
space. Essential dimensions of emotional states are captured by the features of
activation (arousal; measured as an intensity), affect (valence or pleasure; measured
as positive or negative feeling after emotion perception) and power (control;
measured as dominance or submissiveness in emotion expression). The projected
image of the three-dimensional emotional space, on a two-dimensional plane of
activation and affect, is shown in Fig. 1.1. According to the physiology of emotion
production mechanisms, it has been found that the nervous system is stimulated
by the expression of high arousal emotions like anger, happiness and fear. This
phenomenon causes an increased heart beat rate, higher blood pressure, changes
in respiration pattern, greater sub-glottal air pressure in the lungs and dryness of
the mouth. The resulting speech is correspondingly louder, faster and characterized
with strong high-frequency energy, higher average pitch, and wider pitch range
[10]. On the other hand, for the low arousal emotions like sadness, the nervous
system is stimulated causing the decrease in heart beat rate, blood pressure, leading
to increased salivation, slow and low-pitched speech with little high-frequency
energy. Thus, acoustic features such as pitch, energy, timing, voice quality, and
articulation of the speech signal highly correlate with the underlying emotions
[11]. However, distinguishing emotions without any ambiguity, using any one of
the above mentioned dimensions, is a difficult task. For example, both anger and

Fig. 1.1 The distribution of
eight emotions on a two
dimensional emotional plane
of arousal and valence:
Courtesy-[13] (*A* Anger,
C Compassion/Sadness,
D Disgust, *F* Fear,
H Happiness, *N* Neutral,
S Sarcasm, *Su* Surprise)

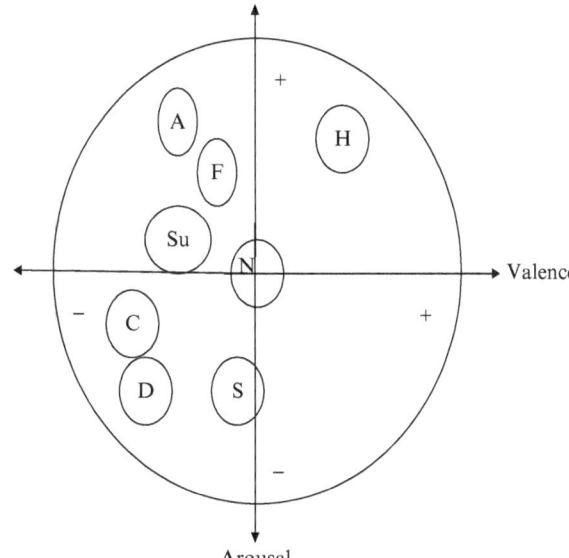

happiness have high activation but they convey different affect (valence or pleasure information). This difference is characterized by using both activation and valence dimensions. Hence, emotion is the complex psycho-physiological experience of an individual's state of mind as interacting with biochemical (internal) and environmental (external) influences. In humans, emotion fundamentally involves physiological arousal, expressive behaviors, and conscious experience. Therefore the three aspects of emotions are psychological (what one is thinking), physiological (what one's body is doing), and expressive (how one reacts) in nature [12].

1.2 *Emotion*: Speech Signal Perspective

Psychological and physiological changes caused due to emotional experience lead to certain actions. Speech is one of the important outcomes of the emotional state of human beings. A speech signal is produced from the contribution of the vocal tract system excited by excitation source signal [14] . Hence, speech specific information may be extracted from vocal tract system and excitation source characteristics. The emotion-specific characteristics of the speech can be attributed to (1) characteristics of the excitation source, (2) the sequence of shapes of the vocal tract system while producing different emotions, (3) supra-segmental characteristics (duration, pitch and energy), and (4) linguistic information.

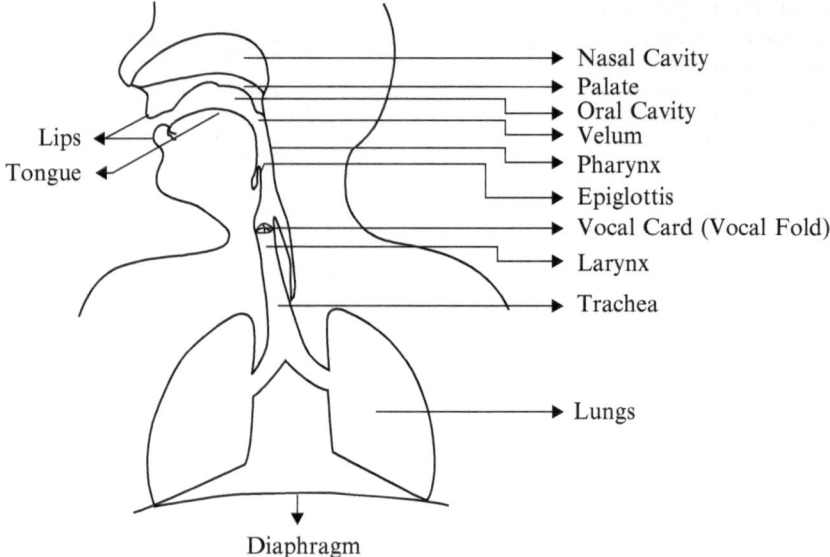

Lips

Tongue

Nasal Cavity
Palate
Oral Cavity
Velum
Pharynx
Epiglottis
Vocal Card (Vocal Fold)
Larynx
Trachea

Lungs

Diaphragm

Fig. 1.2 Human speech production system

1.2.1   Speech Production Mechanism

The air pressure created in the trachea (sub-glottal region) gets released into the vocal tract and nasal cavity, in the form of air puffs. The pressure created beneath the vocal folds causes their vibration, known as glottal activity. The air flow is chopped into a sequence of quasi-periodic pulses through vibration of vocal folds. This sequence of impulse-like excitation gets transformed into different frequency components, while passing through the pharynx, oral, and nasal cavities [15]. The vibration of vocal folds acts as the major source of excitation to the vocal tract. The oral cavity behaves as a time varying resonator to enhance certain frequency components. Movements of articulator organs of the vocal and nasal tracts modulate this quasi-periodic sequence of air puffs into speech, that gets radiated through lips. Vibration of vocal folds leads to the production of voiced sound units. Constriction of a vocal tract at different places, due to the movement of articulators, along with the noisy excitation leads to production of unvoiced sound units. Depending upon the position of various articulators and the phenomenon of voicing or unvoicing, different sound units are produced. The characteristics of glottal activity and vocal tract shapes also play a major role in modulating different emotions, during production of speech. Figure 1.2 shows the speech production system of human beings.

1.2.2 Source Features

During production of speech, vibration of vocal folds provides quasi periodic impulse-like excitation to the vocal tract system. Inverse filtering of a speech signal can remove the vocal tract contribution from the speech signal, and this signal is known as a Linear prediction (LP) residual. In general, the LP residual signal is treated as an approximation of the excitation source signal [14]. Inverse LP analysis removes lower order relations present among the speech samples and retain the higher order relations. The relations present among the adjacent speech samples are treated as lower order relations and those among distant samples are treated as higher order relations. Due to presence of higher order relations, the LP residual signal appears like a random signal and does not contain information beyond the fundamental frequency of speech. The features derived from the LP residual may contain useful information, and it can be used for developing different speech systems. These source features contain supplementary information with respect to the vocal tract features, since the vocal tract system features mainly represent lower order relations present among the speech samples. The features used to represent glottal activity, mainly the vibration of glottal folds, are known as source or excitation source features. These terms are interchangeably used in this book, to represent source information. Figure 1.3 shows the segments of voiced and unvoiced speech signals and their corresponding LP residual signals. The LP residual of the voiced speech segment is observed to be a sequence of impulses corresponding to instants of glottal closure, where the prediction error is maximum. In the case of unvoiced speech, the magnitude of LP residual samples is relatively higher compared to LP residual samples of voiced speech. We propose the source information for characterizing the speech emotions, as the higher order relations present in the LP residual may contain emotion-specific information along with other speech features.

1.2.3 System Features

The sequence of impulse-like excitation, caused due to vocal folds' vibration, acts as a stimulus to the vocal tract system. The vocal tract system can be considered as a cascade of cavities of varying cross sections. The sequences of shapes assumed by the vocal tract, while producing different sound units, are treated as the vocal tract system characteristics of the sound units. During speech production, a vocal tract acts as resonator and emphasizes certain frequency components depending on the shape of the oral cavity. Formants are the resonances of the vocal tract system at a given instance of time. These formants are represented by bandwidth and amplitude, and these parameters are unique to each of the sound units. The sequence of shapes of the vocal tract system also carries emotion-specific information, along with the information related to the sound unit. These features are clearly observed in the

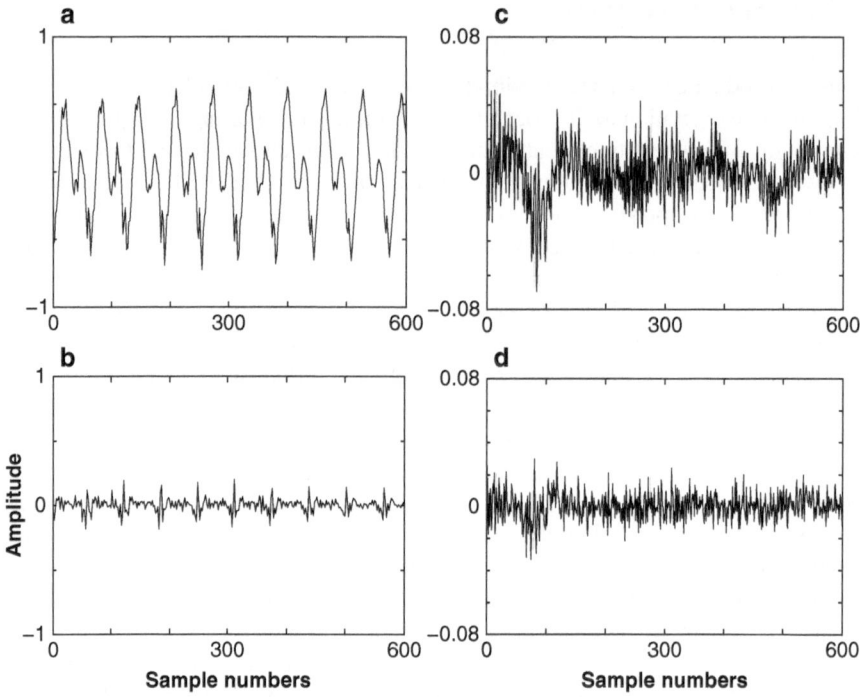

Fig. 1.3 LP residual obtained by inverse LP analysis of the speech. (**a**) Voiced portion of a speech signal, (**b**) corresponding LP residual signal, (**c**) unvoiced portion of the speech signal, and (**d**) corresponding LP residual signal

frequency domain. For frequency domain analysis, a speech signal is segmented into frames of size 20–30 ms, with a frame shift of 10 ms. The spectrum of the speech frame is obtained through Fourier transform. From the magnitude spectrum, the features like linear prediction cepstral coefficients (LPCCs), mel frequency cepstral coefficients (MFCC's), perceptual linear prediction coefficients (PLPCs) and their derivatives are computed to represent vocal tract characteristics [15, 16]. Features extracted from the vocal tract system are generally known as spectral, system or segmental level features. These terms are interchangeably used in this book, to represent vocal tract information.

MFCCs, LPCCs, PLPCs are widely known spectral features used in the literature for various speech tasks. The information about the sequence of shapes of the vocal tract, responsible for producing various is captured through spectral features [15, 16]. The peaks in the spectra are known as formants. They indicate the resonance frequencies of the vocal tract system. Figure 1.4 shows a segment of voiced speech and its LP spectrum. F1, F2, F3, and F4 are formant frequencies.

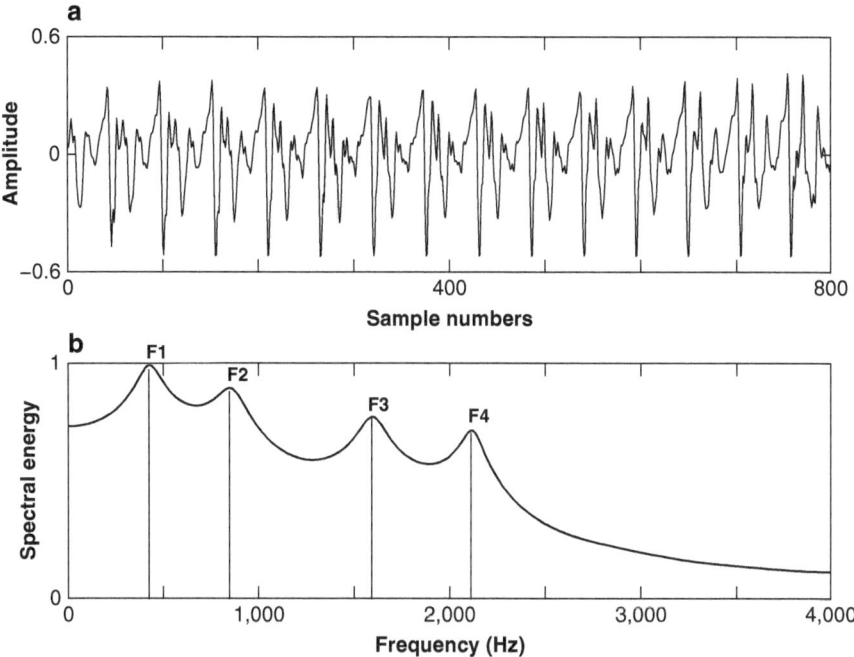

Fig. 1.4 Speech spectrum obtained from frequency domain analysis. (**a**) Voiced speech segment and its, (**b**) magnitude spectrum

1.2.4 Prosodic Features

Speech features extracted from longer speech segments like syllables, words and sentences are known as prosodic features. They represent overall quality of the speech such as rhythm, stress, intonation, loudness, emotion and so on. It is difficult to derive suitable correlates to represent the above mentioned speech quality features. In the literature, pitch, duration, energy and their derivatives are widely used to represent prosodic features. These prosodic features are also known as supra-segmental or long-term features. These terms are interchangeably used in this book, to represent prosodic information. Human beings perceive emotions present in speech, by exploiting the prosodic features, and in this studies, these features are explored for classifying the emotions. Figure 1.5 shows the speech utterance and the corresponding prosodic features.

From the above discussion, the importance to explore the excitation source, vocal tract system and prosodic features, to capture emotion-specific information [17], is observed. This book addresses an issue of *speech emotion recognition* by exploring the above mentioned emotion-specific features, for discriminating the emotions. Terms such as recognition performance, classification performance, and discrimination are used in this book, in the context of emotions, unless specifically mentioned.

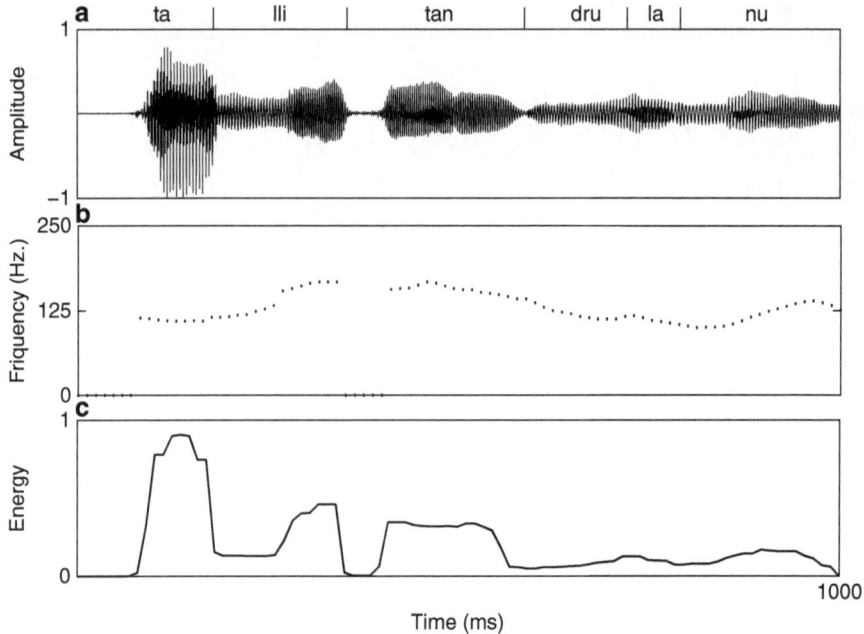

Fig. 1.5 Prosodic features of the speech. (**a**) Speech signal with syllable durations, (**b**) pitch contour, and (**c**) energy contour

1.3   Emotional Speech Databases

An important issue to be considered in evaluating emotional speech systems is the quality of the databases used to develop and assess the performance of the systems [5]. The objectives and methods of collecting speech corpora highly vary according to the motivation behind the development of speech systems. Speech corpora used for developing emotional speech systems can be divided into three types. The important properties of these databases are briefly mentioned in Table 1.1.

1. Actor (Simulated) based emotional speech database
2. Elicited (Induced) emotional speech database
3. Natural emotional speech database

Simulated emotional speech corpora are collected from reasonably experienced and trained theatre or radio artists. Artists are asked to express linguistically neutral sentences in different emotions. Recording is done in different sessions to consider the variations in the degree of expressiveness and physical speech production mechanism of human beings. This is one of the easier and reliable methods of collecting expressive speech databases containing a wide range of emotions. More than 60 % of the databases collected for expressive speech research are of this kind. The emotions collected through simulated means are fully developed in nature,

Table 1.1 Different types of databases used in speech emotion recognition

Type of database	Advantages	Disadvantages
Actor(Simulated)		
E.g.: LDC speech corpus [21], Emo-DB [22], IITKGP-SESC [23]	• Most commonly used • Standardized • Results can be compared easily • Complete range of emotions is available • Wide variety of databases in most of the languages is available	• Acted speech tells how emotions should be portrayed rather than how they are portrayed • Contextual, individualistic and purpose dependent information is absent • Episodic in nature, which is not true in real world situation • Often it is read speech, not spoken
Elicited(Induced)		
E.g.: Wizard of Oz databases, ORESTEIA [24]	• Nearer to the natural databases • Contextual information is present, but it is artificial	• All emotions may not be available. • If the speakers know that they are being recoded, the quality will be artificial
Natural		
E.g.: Call center conversations [25], Cockpit recordings	• They are completely naturally expressed • Mostly useful for real world emotion modeling	• All emotions may not be available • Copyright and privacy issues • Overlapping of utterances • Presence of background noise • Contains multiple and concurrent emotions • Pervasive in nature. So, difficult to model

which are typically intense, and incorporate most of the aspects considered relevant for the emotion [18]. These are also known as *full blown* emotions. Generally, it is found that acted/simulated emotions tend to be more expressed than real ones [5, 19].

Elicited speech corpora are collected by simulating the artificial emotional situation, without the knowledge of the speaker. Speakers are made to involve themselves in emotional conversation with the anchor, where different contextual situations are created by the anchor through conversation to elicit different emotions from the subject, without his/her knowledge. These databases may be more natural than their simulated counterparts, but subjects may not be properly expressive if they know that they are being recorded. Sometimes these databases are recorded by asking the subjects to involve themselves in verbal interaction with a computer whose speech responses are in turn controlled by the human being without the knowledge of the subject [20].

Unlike full blown emotions, natural emotions are mildly expressed. Sometimes, It may be difficult to clearly recognize these emotions. They are also known as *underlying emotions*. Naturally available real world data may be recorded from call center conversations, cockpit recordings during abnormal conditions, a dialog between patient and a doctor and so on. However, it is difficult to find a wide emotion base in this category. Annotation of these emotions is also highly subjective (expert opinion based) and categorization is always debatable. There are also some legal issues, such as privacy and copyright, while using natural speech databases [5, 20]. Table 1.1 briefly explains the advantages and drawbacks of the three types of emotional speech databases.

1.4   Applications of Speech Emotion Recognition

Speech emotion recognition has several applications in day-to-day life. It is particularly useful for enhancing the naturalness in speech-based human machine interaction [23, 26, 27]. Emotion recognition systems may be used in an on-board car driving system, where information about the mental state of a driver may be used to keep him alert during driving. This helps to avoid some accidents, caused by a stressed mental state of the driver [26]. Call center conversation analysis may be helpful behavioral study of call attendants with the customers, and helps to improve the quality of service of a call attendant [25, 28]. Interactive movie[28], story telling [29] and E-tutoring [1] applications would be more practical, if they can adapt themselves to listeners' or students' emotional states. The automatic way to analyze the emotions in speech is useful for indexing and retrieving the audio/video files based on emotions [30]. Medical doctors may use the emotional contents of the patient's speech as a diagnosing tool for various disorders [31]. Emotion analysis of telephone conversation between criminals would help crime investigation departments. Conversation with robotic pets and humanoid partners would be more realistic and enjoyable, if they are able to express and understand emotions like humans [32]. It may also be useful in automatic speech-to-speech translation systems, where the speech in language x is translated into other language y by the machine. Here, both emotion recognition and synthesis are used. The emotions present in the source speech are to be recognized, and the same emotions are to be synthesized in the target speech, as the translated speech is expected to represent the emotional state of the original speaker [5]. In aircraft cockpits, speech recognition systems trained to recognize stressed-speech are used for better performance [33]. Call analysis in emergency services like ambulance and fire brigade may help to evaluate the genuineness of the requests.

1.5   Issues in Speech Emotion Recognition

Some of the important research issues in speech emotion recognition are discussed below in brief.

- The word *emotion* is inherently uncertain and subjective. The term *emotion* has been used with different contextual meanings by different people. It is difficult to define *emotion* objectively, as it is an individual mental state that arises spontaneously rather than through conscious effort. Therefore, there is no common objective definition and agreement on the term *emotion*. This is the fundamental hurdle to proceed with the research [34].

- There are no standard speech corpora for comparing the performance of research approaches used to recognize emotions. Most of the systems processing emotional speech are developed using full blown emotions, but real life emotions are pervasive and underlying in nature. Some databases are recorded using experienced artists, whereas some others are recorded using semi experienced or inexperienced subjects. The research on emotion recognition is limited to five to six emotions, as most of the databases do not provide a wide variety of emotions [21].
- The emotion recognition systems developed using various features may be influenced by the speaker and language specific information. Ideally, speech emotion recognition systems should be speaker and language independent [35].
- An important issue in the development of a speech emotion recognition systems is identification of suitable features that efficiently characterize different emotions [5]. Along with the features, suitable models are to be identified to capture emotion-specific information from the extracted speech features.
- Speech emotion recognition systems should be robust enough to process real-life and noisy speech to identify the emotions.

1.6 Objectives and Scope of the Work

The objective of this work is to study different emotion-specific features from excitation source, vocal tract system and prosodic aspects of speech and suggest best suitable features for the classification of emotions in simulated and natural environments. From the literature, it may be observed that excitation source information is grossly ignored while extracting features for developing speech systems such as: speech synthesis, speech recognition. However, from the speech production point of view, excitation source information is as important as vocal tract information. Therefore, in this work, different source features are extracted from speech signals to study their discriminating characteristics while classifying the emotions. Several evidence are available in the literature, on processing entire speech signal, frame by frame to extract vocal tract system features for performing different speech tasks. On the same lines, In this book, use of basic vocal tract features such as linear prediction cepstral coefficients (LPCCs) and mel frequency cepstral coefficients (MFCCs) is discussed for speech emotion recognition. Prosodic features are treated as effective correlates of speech emotions by a majority of researchers. In this work, specific study has been conducted to analyze the contribution of global and local prosodic features derived from the speech utterances.

1.7   Main Highlights of Research Investigations

- A comprehensive analysis of the literature on *speech emotion recognition* from source, system and prosodic aspects.
- Detailed analysis of different speech corpora used for speech emotion recognition.
- Design and collection of a simulated emotional speech corpus in the Telugu language.
- Proposing various excitation source features for speech emotion recognition.
- Exploring the well known spectral features such as LPCCs, MFCCs and formant frequencies for classifying the speech emotions.
- Proposing global and local prosodic features at the utterance level for emotion recognition.

1.8   Brief Overview of Contributions to This Book

The major contributions of the book include exploring the emotion-specific features from the excitation source, vocal tract system, and prosodic components of speech. The details are discussed in the following sub-sections.

1.8.1   Emotion Recognition Using Excitation Source Information

In general, excitation source information is not thoroughly investigated for developing the speech systems. There is little evidence in the literature, for using this information for different speech tasks [36–42]. In this work, some correlates of excitation source information are explored for speech emotion recognition. Linear prediction (LP) residual, residual phase, epoch parameters, glottal volume velocity (GVV) signal and GVV parameters are explored to study their contribution toward emotion discrimination. From the studies, it is observed that an emotion recognition performance of around 50 % is achieved by using only excitation source information.

1.8.2   Emotion Recognition Using Vocal Tract Information

A conventional block processing approach is widely used in the literature to extract different system features. These features are used to implement different speech systems. In this book, we proposed the same vocal tract system features extracted using block processing for recognition of emotions.

1.8.3 Emotion Recognition Using Prosodic Information

Most of the existing emotion recognition studies have focused on global prosodic features. In this book, local (dynamic) and global (static) prosodic features have been proposed for emotion recognition. These global and local features are extracted from the utterance level and used for emotion recognition.

1.9 Organization of the Book

The evolution of ideas presented in this book is given in the Table 1.2 at the end of this chapter. The chapter wise organization of the book is given below.

- **Chapter 1: The Introduction** introduces the topic of *emotion recognition* from speech. The philosophy of *emotions* with respect to the psychology and speech production aspects is briefly discussed. Different types of emotional speech corpora, contemporary issues in speech emotion recognition, applications, objectives and scope of this book are briefly discussed. Evolution of the ideas of this book work is given at the end of the chapter.
- **Chapter 2: Speech Emotion Recognition: A Review** contains the information about the state-of-the-art literature on emotion-specific features in the context of source, system and prosodic aspects of speech. Detailed study on existing emotional speech databases is also included in the chapter. The chapter concludes with a discussion on unexplored speech features for emotion recognition, and the scope for the present work derived from the review.
- **Chapter 3: Emotion Recognition using Excitation Source Information** discusses the motivation to use excitation source information for speech emotion recognition. Various features derived from the excitation source signal, such as higher order relations present among the samples of LP residual, features of instants of significant excitation and glottal pulse parameters are proposed as important excitation source information for recognizing the speech emotions. Relevant results are presented with appropriate conclusions.
- **Chapter 4: Emotion Recognition using Vocal Tract Information** discuss about the details of development of emotion recognition models using various spectral features derived from the speech signal.
- **Chapter 5: Emotion Recognition using Prosodic Information** presents the details of development of emotion recognition models using prosodic features. Global (static) and local (dynamic) prosodic features extracted from utterance level are proposed in this chapter to discriminate the basic emotions.
- **Chapter 6: Summary and Conclusions** summarizes the contributions of this book along with some important conclusions. This Chapter also provides the extensions to the present work and future directions for improving the performance of emotion recognition models.

Table 1.2 Evolution of ideas presented in this book

Evolution of ideas presented in this book

- Most of the present speech systems are efficient in processing neutral speech
- Incorporation of components of *emotion processing* into today's speech systems makes them natural, robust and pragmatic
- Modeling *emotions* requires a suitable emotional speech corpus

  - Simulated emotional speech corpus in Telugu is collected using radio artists in 8
  - Emotions

- Emotion-specific features need to be identified to develop speech emotion recognition
- Models
- Different speech features are explored toward emotion recognition

  - Excitation source features
  - Vocal tract system features
  - Prosodic features

- Excitation source signal (which appears like noise) is treated not to be of much use for developing any of the speech tasks and grossly ignored by the research community. However excitation signal is essential to stimulate vocal tract for producing speech, and does have significant contribution in the speech production process

  - LP residual, residual phase, epoch parameters, glottal volume velocity (GVV) signal, GVV parameters are used as excitation source features to develop emotion recognition systems

- It is known that the samples of an excitation source signal do not contain linear relations among them. Therefore, auto-associative neural networks, which are known to capture non-linear relations, present among the input vectors, are employed to develop emotion recognition models using source features
- While extracting spectral features, generally a block processing approach is used, where the entire speech signal is processed frame by frame. In this work the same approach has been used in to characterize the emotions
- Gaussian mixture models (GMMs) are well known to capture distributions of input data points. Therefore, GMMs are employed to develop emotion recognition models using spectral features
- Prosodic information is treated to be the major contributor toward characterization of emotions. Static (global) prosodic features are thoroughly explored in the literature for emotion recognition. Dynamic nature of prosodic contours is observed to be more emotion-specific. These static and dynamic prosodic features are extracted at sentence level to characterize the emotions
- Support vector machines (SVMs) are well known for capturing the discriminative information present among the feature vectors. Performance of SVMs is critical on the number of discriminative feature vectors (known as support vectors) rather than the total number of feature vectors. Therefore, SVMs are employed to develop emotion recognition models using prosodic features. There are less number of feature vectors in the case of prosodic analysis of emotions

Chapter 2
Speech Emotion Recognition: A Review

Abstract This chapter presents the literature related to the databases, features, pattern classifiers used for emotion recognition from speech. Different types of emotional databases such as simulated, elicited and natural are critically reviewed from the research point of view. Review of existing emotion recognition systems developed using excitation source, vocal tract system and prosodic features is briefly presented. Basic pattern classification models used for discriminating the emotions are discussed in brief. Finally, the chapter concludes with motivation and scope of the work presented in this book.

2.1 Introduction

Speakers mainly convey their intentions through non-verbal means such as emotions, in the conversation. In addition to the message conveyed through textual contents, the manner in which the words are spoken conveys essential non-linguistic information. The non-linguistic information may be observed through (1) facial expressions in the case of video, (2) expression of emotions in the case of speech, and (3) punctuation in the case of written text. The discussion in this book is confined to emotions or expressions related to speech. Spoken text may have several interpretations, depending on how it is said. For example, the word *OKAY* in English is used to express admiration, disbelief, consent, disinterest or an assertion. Understanding the text alone is not sufficient to interpret the semantics of a spoken utterance. Therefore, it is important that speech systems should be able to process the non-linguistic information such as emotions, along with the message.

Speech is one of the natural modalities of human machine interaction. Today's speech systems may reach human-equivalent performance only when they can process underlying emotions effectively [43]. The purpose of the sophisticated speech systems should not be limited to mere message processing, rather it should understand the underlying intentions of the speaker by detecting their expressions in the speech [1, 44, 45]. In the recent past, processing a speech signal for recognizing

S.R. Krothapalli and S.G. Koolagudi, *Emotion Recognition using Speech Features*,   15
SpringerBriefs in Electrical and Computer Engineering,
DOI 10.1007/978-1-4614-5143-3_2, © Springer Science+Business Media New York 2013

the underlying emotions has emerged as one of the important speech research areas. Embedding the component of *emotion processing* into existing speech systems makes them more natural and effective. Therefore, while developing speech systems (i.e., speech recognition, speaker recognition, speech synthesis and language identification), one should appropriately utilize the knowledge of emotions.

This chapter provides a review of the literature on speech emotion recognition, in view of emotion-specific features extracted from different aspects of speech. The features are broadly classified into three categories namely, excitation source, vocal tract system and prosodic features. Review of some important existing emotional speech corpora is given in Sect. 2.2. Section 2.3 discusses the role of excitation source features for developing different speech systems. The research on recognizing emotions from speech using system features is analyzed in Sect. 2.4. Section 2.5 highlights some of the existing works on speech emotion recognition using prosodic features. Review of the classification models used for speech emotion recognition is briefly discussed in Sect. 2.6. The motivation of the book from the available literature is given in Sect. 2.7. The chapter concludes with Sect. 2.8, by providing the scope of the work.

2.2 Emotional Speech Corpora: A Review

For characterizing the emotions, either for synthesis or for recognition, a suitable emotional speech database is a necessary prerequisite [21]. The design and collection of emotional speech corpora mainly depends on the research goals. For example: a single speaker emotional speech corpus would be enough for the purpose of emotional speech synthesis, whereas recognizing emotions needs a database with multiple speakers and various styles of expressing the emotions. The survey presented in this section critically analyzes the emotional speech databases based on the language, number of emotions and the method of collection. The general issues to be considered while recording the speech corpus are as follows [46].

- The scope of the emotion database both in terms of number of subjects contributing for recording and number of emotions to be recorded is to be decided properly.
- The decision about the nature of the speech as natural or acted, helps to decide the quality and applications of the database.
- Proper contextual information is essential, as naturalness of expressions mainly depends upon the linguistic and general context.
- Labeling of soft emotions present in the speech databases is highly subjective.
- Size of the database used for speech emotion recognition plays an important role in deciding the properties such as scalability, generalizability, and reliability of the developed systems. Most of the existing emotional speech databases used for developing emotion systems are too small in size to capture the influence of speakers, gender, and language for characterizing the emotions [46].

In the literature, emotional speech databases are collected mainly using three different methods. (1) Actors are asked to portray the given emotion in the case of simulated databases. These are the most popular databases used in emotional speech research. (2) Elicited databases are not completely natural, but are recorded under simulated natural situations. (3) Naturalistic databases are recorded form the natural situations. The properties of some important emotional speech corpora being used for emotional speech research are briefly discussed in Table 2.1. From Table, it may be observed that there is a huge disparity among the databases, in terms of language, number of emotions, number of subjects, purpose of corpus collection and methods of database collection.

The set of emotional speech databases, given in Table 2.1, is dominated by the English language, followed by German and Chinese. Very few databases are collected in languages such as: Russian, Dutch, Slovenian, Swedish, Japanese and Spanish. There is no reported reference of an emotional speech database in any of the Indian languages. Among the emotional speech databases, given in Table 2.1, 24 speech corpora are collected for the purpose of recognition and 8 are collected with the intention of synthesis. Subjective listening tests confirm that the average emotion recognition rate in case of any database has not crossed beyond 80%. For full blown emotion subjective listening tests have shown more than 90% of recognition performance. Most of the automatic emotion recognition systems have achieved the recognition performance close to subjective listening tests. About 70% of databases contain only 4–5 basic emotions. Few emotional speech databases contain seven and eight emotions. Most of the existing databases rarely contain uncommon emotions like: antipathy, approval, attention, prohibition, etc. A majority of the databases contain clearly distinguishable emotions such as anger, sadness, happiness and neutral. Since actor based simulated database collection is a straight forward and comparatively easy process, more than half of the databases mentioned in Table 2.1 belong to the category of simulated databases. Sometimes depending upon the need, emotional speech conversations are also recorded from TV shows, and later annotation of emotions is performed by expert artists. From the available emotional speech databases, it is observed that there are no availability of standard, internationally approved database for emotion processing. Recently COCOSDA, The International Committee for the Coordination and Standardization of Speech Databases and Assessment Techniques, which promotes collaboration and information exchange in speech research, has adopted emotional speech as a future priority theme [47]. 'HUMAINE', a group of researchers dedicated to speech emotion recognition, has started the *INTERSPEECH emotion challenge* since 2009, to facilitate- feature, classifier, and open performance comparison for non-prototypical spontaneous emotion recognition. In Indian context, some organizations such as the Linguistic Data Consortium for Indian Languages (LDCIL), Centre for Development of Advanced Computing (CDAC), Tata Institute of Fundamental Research (TIFR), Department of Information Technology (DIT-Technology Development for Indian Languages) are contributing toward speech data collection. However, they confined themselves to collect speech corpora in different Indian languages in the context of speech recognition/synthesis and speaker recognition tasks only.

Table 2.1 Literature survey of speech databases used for emotion processing

Sl. no.	Emotions	Number of speakers	Type of database	Purpose and approach	References
English emotional speech corpora					
01	Depression and neutral (02)	22 patients and 19 healthy persons	Simulated	Recognition. Prosody variations are analyzed with respect to the speech samples of depressed and healthy people	[48]
02	Anger, disgust, fear, joy, neutral, sadness and surprise (07)	Eight actors (two per language) (two per language)	Simulated	Synthesis. Speech in four languages (English, Slovenian, Spanish, and French) is recorded	[49]
03	Anger, boredom, joy, and surprise (04)	51 children	Elicited	Recognition. Recorded at the university of Maribor, in German and English	[50]
04	Anger, fear, happiness, neutral, and sadness (05)	40 native speakers	Natural	Recognition. Two broad domains of emotions are proposed based on prosodic features	[51]
05	Different natural emotions	125 TV artists	Natural	Recognition. It is known as Belfast natural database and is used for several emotion processing applications	[52]
06	Anger, boredom, fear, happiness, neutral, and sadness (06)	Single actor	Simulated	Synthesis. F_0, duration and energy are modeled for synthesizing the emotions	[53]
07	Depression and neutral (02)	70 patients 40 healthy persons	Natural	Recognition. F_0, amplitude modulation, formants, power distribution are used to analyze depressed and suicidal speech	[31]

	German emotional speech corpora				
08	Depression and neutral (02)	Different native speakers	Elicited	Recognition	[54]
09	Negative and positive (02)	Customers and call attendants	Natural	Recognition. Call center conversations are recorded	[25]
10	Annoyance, shock and stress (03)	29 native speakers	Elicited	Recognition	[24]
11	Hot anger, cold anger, happiness, neutral, and sadness (05), 40 utterances per emotion are recorded.	29 native speakers	Elicited	Recognition. Dimensional analysis of emotions is performed using F0 parameters	[55]
12	Anger, fear, neutral, and sadness (04)	Different native speakers	Simulated	Recognition. Prosodic, spectral and verbal cues are used for emotion recognition	[56]
13	Five stress levels (05)	6 soldiers	Natural	Recognition	[57]
14	Two task load stress conditions and two normal stress conditions (02)	100 native speakers	Natural	Recognition. Effects of stress and load on speech rate, F0, energy, and spectral parameters are studied. The databases are recorded in English and German	[58]
15	Approval, attention, and prohibition (03)	12 native speakers	Natural	Recognition. Pitch and broad spectral shapes are used to classify adult-directed and infant-directed emotional speech (BabyEars). The databases are recorded in English and German	[59]

(continued)

Table 2.1 (continued)

Sl. no.	Emotions	Number of speakers	Type of database	Purpose and approach	References
Japanese emotional speech corpus					
16	Anger, happiness, neutral, sadness (04), 112 utterances per emotion are recorded.	Single actress	Simulated	Recognition. Speech prosody, vowel articulation and spectral energy distribution are used to analyze four emotions	[60]
17	Anger, Boredom, disgust, fear, joy, neutral, and sadness (07)	Ten actors	Simulated	Synthesis	[61]
18	Different elicited emotions are recorded	51 school children (21M+30F)	Elicited	Recognition. Children are asked to spontaneously react with Sony AIBO pet robot. Around 9.5 h of effective emotional expressions of children are recorded	[62]
19	Anger, Boredom, disgust, fear, joy, neutral, and sadness (07)	Ten actors (5M+5F)	Simulated	Recognition. About 800 utterances are recorded using 10 neutral German sentences	[22]
20	Soft, modal, and loud (03)	Single actor	Simulated	Synthesis. Di-phone based approach is used for emotional speech synthesis	[63]
21	Anger, Boredom, disgust, and worry (04)	Six native speakers	Simulated	Recognition. Affective bursts and short emotional non-speech segments are analyzed for discriminating the emotions	[64]
22	Two emotions for each emotional dimension are recorded. (1) Activation (calm-excited), (2) Valence (positive-negative), and (3) Dominance (weak-strong)	104 native speakers (44M+60F)	Natural	Recognition. Twelve hours of audio visual-recording is done using TV talk show *Vera am Mittag* in German. Emotion annotation is done based on activation, valence, and dominance dimensions	[65]

Chinese emotional speech corpora

23	Antipathy, anger, fear, happiness, sadness, and surprise (06)	Two actors	Simulated	Recognition	[66]
24	Anger, disgust, fear, joy, sadness, and surprise (06), 60 Utterances per emotion per speaker are recorded	12 actors	Simulated	Recognition. Log frequency power coefficients are proposed for emotion recognition using HMMs	[67]
25	Anger, happiness, neutral, and sadness (04), 721 short utterances per emotion are recorded	Native TV actors	Simulated	Recognition	[68]
26	Anger, fear, joy, neutral and sadness (05), 288 sentences per emotion are recorded	Nine native speakers	Elicited	Recognition. Phonation, articulation and prosody are used to classify four emotions	[69]

Spanish emotional speech corpora

27	Desire, disgust, fear, fury (anger), joy, sadness, and surprise (07)	Eight actors (4M+4F)	Simulated	Synthesis. Acoustic modeling of Spanish emotions is studied. Rules are used to identify significant behavior of emotional parameters	[70]
28	Anger, disgust, happiness, and sadness (04), 2,000 phones per emotion are considered	Single actor	Simulated	Synthesis. Pitch, tempo, and stress are used for emotion synthesis	[71]

(continued)

Table 2.1 (continued)

Sl. no.	Emotions	Number of speakers	Type of database	Purpose and approach	References
29	Anger, joy, and sadness (03)	Two native speakers	Simulated	Synthesis Concatenative synthesis approach is used	[72]
	Russian emotional speech corpus				
30	Anger, fear, happiness, neutral, sadness, and surprise (06), ten sentences are recorded per emotion in different sessions	61 Native speakers	Simulated	Recognition. This database is used for both language and speech processing applications (RUSSLANA)	[73]
	Swedish emotional speech corpus				
31	Happiness and neutral (02)	Single native speaker	Simulated	Synthesis. Variations in articulatory parameters are used for uttering Swedish vowels in two emotions	[74]
	Italian emotional speech corpus				
32	Anger, disgust, fear, joy, sadness, and surprise (06)	Single native speaker	Simulated	Synthesis	[75]

From the above mentioned survey, it is observed that there is a need of an emotional speech corpus in Indian languages, to support the research on speech emotion recognition and synthesis, in the context of Indian languages. The database may contain seven and eight common emotions. Though it is actor based simulated speech corpus, it should contain underlying emotions, rather than full blown emotions, so that the expression of emotions is more real and natural. The database has to contain sufficient variability in terms of number of speakers, gender, and sessions of recording. The text prompts used for recording are to be neutral in nature, so that linguistic contents do not influence the expressiveness of emotions.

2.3 Excitation Source Features: A Review

Speech features derived from the excitation source signal are known as source features. The excitation source signal is obtained from speech, after suppressing vocal tract (VT) characteristics. This is achieved by, first predicting the VT information using filter coefficients (linear prediction coefficients (LPCs)) from the speech signal, and then separating it by inverse filter formulation. The resulting signal is known as a *linear prediction residual*, and it contains mostly the information about the excitation source [76]. In this book, features derived from the LP residual are referred to as the excitation source or simply source features. The sub-segmental analysis of the speech signal is aimed at studying characteristics of the glottal pulse, open and closed phases of glottis, strength of the excitation and so on. The characteristics of glottal activity, specific to the emotions may be estimated using excitation source features. The LP residual signal and glottal volume velocity (GVV) signal are explored in the literature as the correlates of excitation source information [77]. In the literature, very few attempts have been made to explore the excitation source information for any of the speech tasks. The reasons may be

1. Popularity of spectral features
2. The excitation signal (LP residual) obtained from LP analysis is viewed mostly as an error signal [78] or unpredictable components of the predicted speech signal.
3. The LP residual basically contains the higher order relations, and capturing these higher order relations is not well known [79].

It may be difficult to parameterize the LP residual signal, however it contains valid information as it provides primary excitation to the vocal tract system, while producing speech. LP residual signal basically contains the higher order correlations among its samples [80], as the first and second order correlations are filtered out during LP analysis. These higher order correlations may be captured to some extent, by using the features like strength of excitation, characteristics of glottal volume velocity waveform, shapes of the glottal pulse, characteristics of open and closed phases of glottis and so on.

The existing studies based on excitation source features of speech have clearly demonstrated that excitation source information contains all flavors of speech

Table 2.2 Literature review on use of excitation source information for various speech tasks

Sl. no.	Features	Purpose and approach	References
01	LP residual energy	Vowel and speaker recognition	[36]
02	LP residual	Instants of significant excitation are determined	[37]
03	Higher order relations among LP residual samples	Categorizing audio documents	[38]
04	LP residual	Speech enhancement in multi-speaker environment	[39]
05	LP residual	Characterizing loudness, lombard effect, speaking rate, and laughter segments	[86]
06	Glottal excitation signal	Analyzing the relation between emotional state of the speaker and glottal activity	[41]
07	Glottal excitation signal	To analyze emotion related disorders	[41]
08	Excitation source signal	To discriminate emotions in continuous speech	[42]

such as message, speaker, language, and emotion-specific information. Probably, the available excitation source features may not compete with well established spectral and prosodic features. Some of the important references regarding the use of excitation information in developing different speech systems are given below. Pitch information extracted from the LP residual signal is successfully used in [81], for speaker recognition. LP residual energy is used in [36], for vowel and speaker recognition. Cepstral features derived from the LP residual signal are used in [82], for capturing the speaker specific information. The combination of features derived from the LP residual and LP residual cepstrum has been used to minimize the equal error rate in case of speaker recognition [83]. By processing LP residual signal using Hilbert envelope and group delay function, the instants of significant excitation are accurately determined [37]. The higher order relations among the samples of LP residual are also used for categorizing different audio documents like: sports, news, cartoons, music in noisy and clean environments [38]. The instants of significant excitation obtained from the LP residual signal during the production of voiced speech are used to determine the relative delays between the speech segments of different speakers in a multi-speaker environment, and they are further used to enhance the speech of individual speakers [39]. The epoch (instants of glottal closure) properties of LP residual are exploited in [84], for enhancing the reverberant speech. The parameters extracted from the excitation source signal at the epoch locations are exploited for analyzing loudness, lombard effect, speaking rate and detecting the laughter segments from the speech [40, 85, 86]. Table 2.2 shows some of the important achievements in speech research using excitation source information.

From the available literature, it is clear that the excitation source information can be used equally well to develop any speech systems compared to spectral and prosodic features. Excitation source information is not exhaustively and systematically explored for speech emotion recognition. The excitation source signal may also contain the emotion-specific information, in the form of higher order relations among linear prediction (LP) residual samples, parameters of instants of significant excitation, parameters of glottal pulse and so on. Hence, there is a scope for conducting the detailed and systematic study on excitation source information for characterizing the emotions.

2.4 Vocal Tract System Features: A Review

Generally, a speech segment of length 20–30 ms is used to extract vocal tract system features. It is known that vocal tract characteristics are well reflected in frequency domain analysis of speech signals. The Fourier transform of a speech frame gives its short time spectrum. Features like formants, their bandwidths, spectral energy and slope may be observed from the spectrum. The cepstrum of a speech frame is obtained by taking the Fourier transform on the log magnitude spectrum [15]. MFCCs (Mel frequency cepstral coefficients) and LPCCs (Linear prediction cepstral coefficients) are the common features derived from the cepstral domain that represent vocal tract information. These vocal tract features are also known as segmental, spectral or system features. In general spectral features are treated as the strong correlates of varying shapes of the vocal tract and the rate of change in the articulator movements [16]. The emotion-specific information present in the sequence of shapes of vocal tract may be responsible for producing different sound units in different emotions. MFCCs, LPCCs, perceptual linear prediction coefficients (PLPCs), and formant features are some of the widely known system features used in the literature [21].

Generally, spectral features have been successfully used for various speech tasks including development of speech and speaker recognition systems. Some of the important works on emotion recognition using spectral features are discussed below. MFCC features are used in [87], to distinguish speech and non-speech (music) information. It has been observed that the lower order MFCC features carry phoneme information, whereas higher order features contain non-speech information. A combination of MFCCs, LPCCs, RASTA PLP coefficients and log frequency power coefficients (LFPCs) is proposed as the feature set, to classify anger, boredom, happiness, neutral and sadness emotions in Mandarin [88, 89]. Log frequency power coefficients (LFPC) are used to represent the emotion-specific information in [10], for classifying six emotions. A four stage ergodic hidden Markov model (HMM) is used as a classifier to accomplish this task. Performance of LFPC parameters is compared with conventional LPCC and MFCC features, and observed that LFPCs perform slightly better [10, 90]. The MFCC features extracted from lower frequency components (20–300 Hz) of the speech signal are proposed to model pitch variation. These are known as MFCC-low features and used to recognize emotions in Swedish and English emotional speech databases.

Table 2.3 Literature review on emotion recognition using vocal tract system features

Speech emotion research using vocal tract system features			
Sl. no.	Features	Purpose and approach	References
01	MFCC features	Discrimination of speech and music. Higher order MFCCs contain more music specific information and a lower number of MFCCs contain more speech specific information	[87]
02	MFCCs, LPCCs RASTA PLP coefficients, log frequency power coefficients	Classification of four emotions in the Mandarin language. Anger, happiness, neutral and sadness emotions are considered in this study	[88, 89]
03	Combination of MFCCs and MFCC-low features	Emotion classification using Swedish and English emotional speech databases	[91]
04	MFCC features from consonant, stressed and unstressed vowels (class-level MFCCs)	Emotion classification on English LDC and Emo-DB databases	[92]
05	Spectral features obtained using Fourier and Chirp transformations	Modeling human emotional states under stress	[93]

MFCC-low features are reported to perform better than pitch features in the case of emotion recognition [91]. Mel-frequency cepstral coefficients computed over three phoneme classes namely: stressed vowels, unstressed vowels and consonants are used for speaker-independent emotion recognition. These features are referred to as class-level spectral features. Classification accuracies are observed to be consistently higher for class-level spectral features than prosodic or utterance-level spectral features. The combination of class-level features with prosodic features improved the emotion recognition performance. Further, results showed that spectral features computed from consonant regions contain more emotion-specific information than either stressed or unstressed vowel features. It is also reported in this work that the average emotion recognition performance is proportional to the length of the utterance [92]. In [93] spectra of vowel segments obtained using Fourier and Chirp transforms are analyzed for emotion classification and observed that the higher frequency regions of speech are suitable for characterizing stressed speech. These features are used to model the emotional state of a stressed person. Some of the efforts on the use of system features for speech emotion recognition are given in Table 2.3. From the references mentioned in Table 2.3, it is observed that, in most of the cases, spectral features are extracted through a conventional block processing approach, wherein the entire speech signal is processed frame by frame, considering the frame size of 20 ms, and a shift of 10 ms.

2.5 Prosodic Features: A Review

Human beings impose duration, intonation, and intensity patterns on the sequence of sound units, while producing speech. The incorporation of these prosody constraints (duration, intonation, and intensity) makes human speech natural. Lack of prosody knowledge can easily be perceived from the speech. Prosody can be viewed as speech features associated with larger units such as syllables, words, phrases and sentences. Consequently, prosody is often considered as supra-segmental information. The prosody appears to structure the flow of speech. The prosody is represented acoustically by the patterns of duration, intonation (F_0 contour), and energy. They normally represent the perceptual speech properties such as: intonation and energy, that are normally used by human beings to perform various speech tasks including emotion recognition [94, 95]. In the literature, mainly, pitch, energy, duration and their derivatives are used as the acoustic correlates of prosodic features [96, 97]. Human emotional expressiveness (i.e. emotionally excited behavior of articulators) can be captured through prosodic features. The prosody can be distinguished at four principal levels of manifestation [95]. They are at (a) Linguistic intention level, (b) articulatory level, (c) acoustic realization level and (d) perceptual level.

At the linguistic level, prosody refers to relating different linguistic elements of an utterance to bring out required naturalness. For example, the linguistic distinctions that can be communicated through distinction between question and statement, or the semantic emphasis on an element. At the articulatory level, prosody is physically manifested as a series of articulatory movements. Thus, prosody manifestations typically include variations in the amplitudes of articulatory movements as well as the variations in air pressure. Muscle activity in the respiratory system as well as along the vocal tract, leads to radiation of sound waves. The acoustic realization of prosody can be observed and quantified using the analysis of acoustic parameters such as fundamental frequency (F_0), intensity, and duration. For example, stressed syllables have higher fundamental frequency, greater amplitude and longer duration than unstressed syllables. At the perception level, speech sound waves enter the ears of the listener who derives the linguistic and paralinguistic information from prosody via perceptual processing. During perception, prosody can be expressed in terms of subjective experience of the listener, such as pauses, length, melody and loudness of the perceived speech. It is difficult to process or analyze the prosody through speech production or perception mechanisms. Hence the acoustic properties of speech are exploited for analyzing the prosody.

In the literature, prosodic features such as energy, duration, pitch and their derivatives are treated as good correlates of emotions [25, 34, 67, 98]. Features such as minimum, maximum, mean, variance, range and standard deviation of energy, and similar features of pitch are used as important prosodic information sources for discriminating the emotions [99, 100]. Some studies [100, 101] have also tried to measure the steepness of the F0 contour during rise and falls, articulation rate, number and duration of pauses for characterizing the emotions.

Prosodic features extracted from the smaller linguistic units like syllables and at the level of consonants and vowels are also used for analyzing the emotions [100]. The importance of prosodic contour trends in the context of different emotions is discussed in [102, 103]. Peaks and troughs in the profiles of fundamental frequency and intensity, durations of pauses and bursts are proposed for identifying four emotions, namely fear, anger, sadness and joy. Around 55% of average emotion recognition performance is reported using discriminant analysis [104]. The sequences of frame-wise prosodic features, extracted from longer speech segments such as words and phrases are also used to categorize the emotions present in the speech [67]. F_0 information is analyzed for emotion classification and it is reported that minimum, maximum and median values of F_0 and slopes of F_0 contours are emotion salient features. Around 80% of emotion recognition accuracy is achieved, using proposed F_0 features with a K-nearest neighbor classifier [27]. Short time supra-segmental features such as pitch, energy, formant locations and their bandwidths, dynamics of pitch, energy and formant contours, speaking rate are used for analyzing the emotions [1]. The complex relations between pitch, duration and energy parameters are exploited in [72] for detecting the speech emotions. Table 2.4 shows some of the other important and recent works on speech emotion recognition using prosodic features.

From the literature, it is observed that most of the speech emotion recognition studies are carried out using utterance level static (global) prosodic features [23, 34, 67, 72, 98, 112]. Very few attempts have explored the dynamic behavior of prosodic patterns (local) for analyzing speech emotions [104, 113]. Elementary prosodic analysis of speech utterances is carried out in [114], at sentence, word, and syllable levels, using only the first order statistics of basic prosodic parameters. In this context, it is important to study the contribution of static and dynamic (i.e., global and local) prosodic features extracted from sentence, word and syllable segments toward emotion recognition. None of the existing studies has explored the speech segments with respect to their positional information for identifying the emotions. The approach of recognizing emotions from the shorter speech segments may further be helpful for real time emotion verification.

2.6   Classification Models

In the literature, several pattern classifiers are explored for developing speech systems like speech recognition, speaker recognition, emotion classification, speaker verification and so on. However, the justification for choosing a particular classifier to the specific speech task is not provided in many instances. Most of the times appropriately some classifiers are chosen. Few times a particular one is chosen among the available alternatives based on experimental evaluation. Wang et al. have conducted the studies on the performance of various classification tools

Table 2.4 Literature review on emotion recognition using prosodic features

Sl. no.	Features	Purpose and approach	References
Speech emotion research using prosodic features			
01	Initially 86 prosodic features are used, later best 6 features are chosen from the list	Identification of emotions in the Basque language. Around 92% emotion recognition performance is achieved using GMMs	[105]
02	35 dimensional prosodic feature vectors including pitch, energy, and duration are used	Classification of seven emotions of the Berlin emotional speech corpus. Around 51% emotion recognition results are obtained for speaker independent cases using neural networks	[106]
03	Pitch and power based features are extracted from frame, syllable, and word levels	Recognizing emotions in Mandarin. Combination of features from frame, syllable and word level yielded 90% emotion recognition performance	[107]
04	Duration, energy, and pitch based features	Recognizing emotions in the Mandarin language. Sequential forward selection (SFS) is used to select best features from the pool of prosodic features. Emotion classification studies are conducted on a multi-speaker multi-lingual database. Modular neural networks are used as classifiers	[108]
05	Eight static prosodic features and voice quality features	Classification of six emotions (anger, anxiety, boredom, happiness, neutral, and sadness) from the Berlin emotional speech corpus. Speaker independent emotion classification is performed using Bayesian classifiers	[109]
06	Energy, pitch and duration based features	Classification of six emotions from Mandarin language. Around 88% of average emotion recognition rate is reported using SVM and genetic algorithms	[110]
07	Prosody and voice quality based features	Classification of four emotions namely anger, joy, neutral, and sadness from the Mandarin language. Around 76% emotion recognition performance is reported using support vector machines (SVMs)	[111]

as applied to speech emotion recognition [115]. In general, pattern recognizers used for speech emotion classification can be categorized into two broad types namely

1. Linear classifiers and
2. Non-linear classifiers

A linear classifier performs the classification by making a classification decision based on the value of a linear combination of the object characteristics. These characteristics are also known as feature values and are typically presented to the classifier in the form of an array called a feature vector. If the input feature vector to the classifier is a real vector $\mathbf{x}$, then the output score is given by $y = f(\mathbf{w} \cdot \mathbf{x}) = f\left(\sum_j w_j x_j\right)$, where $\mathbf{w}$ is a real vector of weights and f is a function that converts the dot product of the two vectors into the desired output. The weight vector $\mathbf{w}$ is learned from a set of labeled training samples. j is the dimension of the feature vectors. Often f is a simple function that maps all values above a certain threshold to the first class and all other values to the second class. A more complex f might give the probability that an item belongs to a certain class.

A non-linear weighted combination of object characteristics is used to develop non-linear classifiers. During implementation, proper selection of a kernel function makes the classifier either linear, or non-linear (Gaussian, polynomial, hyperbolic, etc.). In addition, each kernel function may take one or more parameters that would need to be set. Determining an optimal kernel function and parameter set for a given classification problem is not a solved problem. There are only useful heuristics to reach satisfying performance. While adopting the classifiers to the specific problem, one should be aware of the facts that non-linear classifiers have a higher risk of over-fitting, since they have more dimensions of freedom. On the other hand a linear classifier has less degree of freedom to fit the data points, and it severely fails in the case of data that is not linearly separable.

Determination of classifier parameters for linear classifiers is done by two broad methods. The first method uses probability density functions and the second method works on discriminative properties of the data points. Some important examples of classifiers using probability density functions are linear discriminant analysis, Fischer's linear discriminant analysis, Naive Bayes classifier, principal component analysis and so on. Important examples of linear classifiers working on discrimination of feature vectors are logistic regression, least square methods, perceptron algorithm, linear support vector machines, Kozinec's algorithm and so on. Discriminative classifiers perform mainly on the principle of non-probabilistic binary classification by adopting supervised learning, whereas probabilistic classifiers adopt unsupervised learning algorithms. Common non-linear classification tools used for general pattern recognition are Gaussian mixture models, hidden Markov models, soft (non-linear) SVMs (Support Vector Machines), neural networks, polynomial classifiers, universal approximators, and decision trees. Types of the pattern classifiers mainly used for speech emotion recognition are given in Fig. 2.1.

Use of classifiers mainly depends upon the nature of data. If the nature of data is known before, then deciding on the type of classifier would be an easier task. Linear

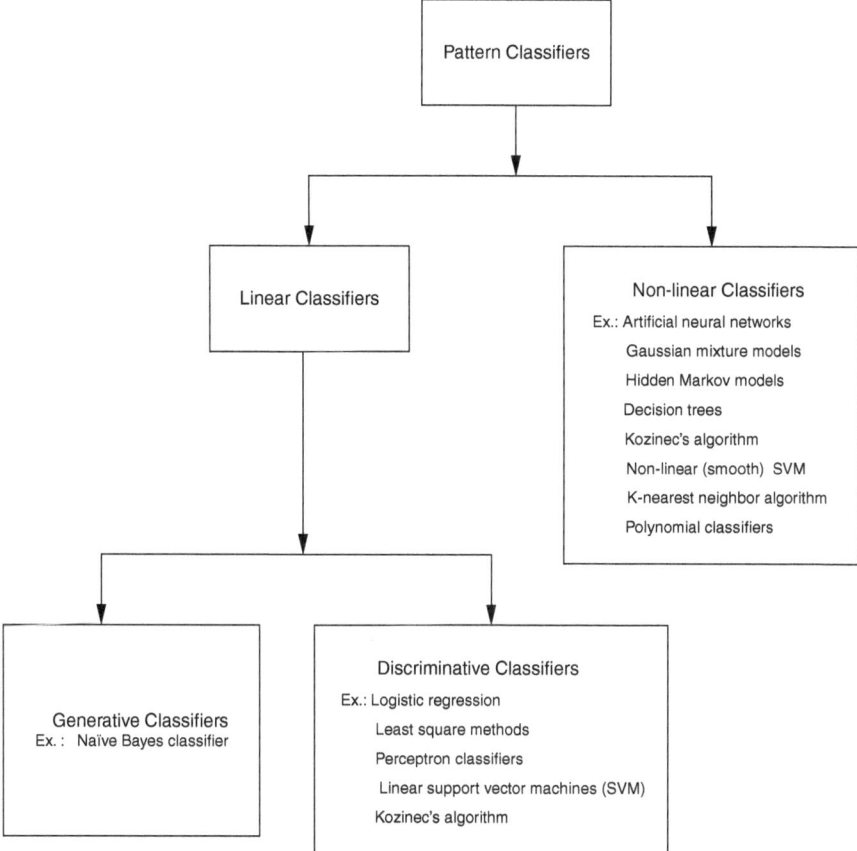

Fig. 2.1   Types of classifiers used for speech emotion recognition

classifiers would classify the features better and faster, if they are clearly, linearly separable. Supervised learning would be helpful, if training data set is properly labeled. Feature vectors not linearly separable would need non-linear classifiers for classification. In most of the real world situations, the nature of the data is rarely known. Therefore, researchers use non-linear classifiers always at the cost of complexity and computational time. Table 2.5 provides the list of classifiers used for speech emotion recognition. From Table 2.5, it may be observed that the majority of the speech emotion recognition tasks have employed non-linear classifiers. Artificial neural networks (ANN), Gaussian mixture models (GMM), and support vector machines (SVM) have been widely used in emotion recognition research. ANNs are known to capture non-linear relations among the feature vectors. GMMs are expected to capture the distribution of input feature space and probabilistically take the decision related to the class of the unknown feature vector. SVMs are discriminative classifiers. Their performance basically depends upon the number of feature vectors with discriminative properties, known as support vectors, rather

Table 2.5 Literature on use of different classifiers for speech emotion recognition task

Sl. no.	Classifiers	Features	References
01	Gaussian mixture models	Prosodic	[26, 59, 91, 115, 116]
	(GMM)	Spectral	[26, 59, 87, 105, 115, 116]
02	Support vector machines	Prosodic	[26, 105, 107, 110, 111, 116, 117]
	(SVM)	Spectral	[26, 107, 116, 117]
03	Artificial neural networks	Prosodic	[26, 118–122],
	(ANN)		[28, 108, 115, 123]
		Spectral	[26, 28, 115, 118, 119, 123, 124]
04	k-Nearest neighbor	Prosodic	[27, 88, 98, 115, 117]
	classifier	Spectral	[115, 117, 125, 126]
05	Bayes classifier	Prosodic	[98, 109, 110, 122]
		Spectral	[109]
06	Linear discriminant	Prosodic	[60, 104, 112]
	analysis with Gaussian	Spectral	[25, 60, 112, 126]
	probability distribution		
07	Hidden Markov models	Prosodic	[67, 92, 122, 127]
	(HMM)	Spectral	[10, 67, 90, 127]

than the total number of input feature vectors. ANNs are mostly used to capture emotion-specific information present in the feature vectors in the form of non-linear higher order relations. Therefore ANN models are suitable for developing the emotion recognition systems using excitation source information. GMMs are used as classifiers, when the number of feature vectors is large enough to capture the proper distribution. For example in the case of spectral features GMMs are suitable for emotion recognition. Systems to be developed using frame-wise spectral features may perform better with GMMs. SVMs are used to develop emotion recognition models using prosodic features, where the number of feature vectors is less, and they have sufficient discriminative characteristics.

2.7 Motivation for the Present Work

The objective of this work is to develop a suitable simulated emotional speech database to promote the research on processing emotions from speech in the context of Indian languages and exploring various emotion-specific features of speech for characterizing the speech emotions. Excitation source, vocal tract system and prosodic information represent three different aspects of speech. Therefore emotion-specific features can be explored in these three broad categories. From the existing work presented in Sect. 2.3, it is observed that no systematic study is carried out on speech emotion recognition using excitation source features. However, excitation source features have been used successfully for some speech tasks [36–38, 42, 86]. Therefore, in this work we are exploring different excitation source features for recognizing speech emotions.

From the literature related to emotion recognition using system features, it is observed that spectral features are mostly extracted through a conventional block processing approach. In this approach feature vectors are extracted from the entire speech signal using overlapped frames. High amplitude regions of spectrum are robust in the case of speech with background disturbances. Therefore, formant features extracted from speech are explored for emotion recognition in combination with other spectral features.

Considering the literature provided in Sect. 2.5, one can observe that most of the existing works on prosodic features mainly exploited their gross statistics at the utterance level such as maximum, minimum, mean, and standard deviation of the feature set for recognizing the emotions. However the variations in prosodic features with respect to time are not explored. In reality, the dynamics of prosodic parameters (i.e., local or fine variations in prosodic parameters with respect to time) are crucial in analyzing the manifestation of the emotions at the suprasegmental level. With this motivation, dynamics of prosodic features are explored along with static features for characterizing the emotions.

2.8 Summary of the Literature and Scope for the Present Work

From the literature, it is observed that there is no proper emotional speech corpus in any of the Indian languages for carrying out the research on emotional speech processing. A real life emotional speech database is also not available in the context of Indian languages. It is also observed from the literature that excitation source information is not thoroughly investigated for the purpose of emotion recognition task. Most of the researchers have used frame-wise spectral features extracted from entire utterance for speech emotion classification. Most of the existing emotion recognition systems are developed using only gross prosodic features extracted from the entire utterances. Dynamics of prosodic patterns and emotion-specific prosody are not explored in view of recognizing the emotions. From this summary of the available work, the scope of this book may be outlined as follows.

- Design and collection of a simulated Telugu emotional speech database from the artists.
- The excitation source features such as LP residual signal, LP residual phase, epoch parameters such as strength of epochs, instantaneous frequency, sharpness of epochs, slope of strength of epochs, glottal volume velocity (GVV) waveform, GVV parameters, dynamics of epoch parameters at syllable and utterance levels, may be used as features for speech emotion recognition.
- Exploring basic spectral features such as LPCCs and MFCCs, extracted from entire speech utterances through block processing approach for recognizing the emotions.

- Static (global) and dynamic (local) features of prosodic contours may be explored for emotion classification.
- Different linear and non-linear pattern classifiers may be explored to study their suitability for emotion classification tasks. AANNs, GMMs, and SVMs are identified as suitable models for developing emotion recognition systems, for classifying the emotions using excitation source, spectral, and prosodic features respectively.

Chapter 3
Emotion Recognition Using Excitation Source Information

Abstract This chapter provides the details of various excitation source features used for recognizing the emotions. The motivation to explore the excitation source information for emotion recognition is illustrated by demonstrating the speech files with source information alone. Details of extraction of proposed excitation source features ((i) Sequence of LP residual samples, (ii) LP residual phase, (iii) Epoch parameters and (iv) Glottal pulse parameters) are given. Two emotional speech databases are introduced to validate the proposed excitation source features. Functionality of classification models such as auto-associative neural networks and support vector machines is briefly explained. Finally, recognition performance using the proposed excitation source features is analyzed in detail.

3.1 Introduction

In the previous two chapters, *speech emotion recognition* is introduced as one of the important research areas and the related work is also discussed. This chapter basically deals with the use of excitation source information for recognizing underlying emotions from speech utterances [128]. Among the different speech information sources, excitation source information is treated almost like a noise and not contain information beyond the fundamental frequency of speech (because it mostly contains unpredictable part of the speech), and grossly ignored by speech research community[128]. Some of the speech systems developed using excitation source features are discussed in the second chapter. However, systematic study has not been carried out on speech emotion recognition using excitation information. The linear prediction (LP) residual represents the prediction error in the LP analysis of speech, and it is considered as the excitation signal to the vocal tract system, while producing the speech. In this chapter, we have explored the features extracted from the LP residual, epochs and GVV waveform for classifying the speech emotions. These features are referred to as excitation source or simply source features.

S.R. Krothapalli and S.G. Koolagudi, *Emotion Recognition using Speech Features*,   35
SpringerBriefs in Electrical and Computer Engineering,
DOI 10.1007/978-1-4614-5143-3_3, © Springer Science+Business Media New York 2013

Epoch is an event representing the instant of glottal closure during the production of voiced speech. The glottal volume velocity (GVV) signal represents the airflow pattern through the glottis.

The chapter is organized as follows. Motivation for exploring the excitation source features is discussed in Sect. 3.2. The emotional speech corpora used in this work are explained in brief in Sect. 3.3. Section 3.4 contains the detailed explanation extraction of various excitation source features, proposed in this chapter. Section 3.5 discusses the details of the classification models used in this work for classifying the emotions. Section 3.6 discusses the results of speech emotion recognition using the proposed excitation source information. Section 3.7 summarizes the topics discussed in this chapter.

3.2   Motivation

From the speech production mechanism, one can observe the speech as the joint contribution of both vocal tract system and excitation source [15]. This indicates that the information present in the speech such as: message, language, speaker and emotion is present in both excitation source and vocal tract characteristics. To strengthen the argument, perceptual study has been carried out to analyze the presence of emotion-specific information in (1) excitation source, (2) the response of vocal tract system and (3) combination of both. For this study, five Telugu sentences uttered in eight different emotions by a female radio artist are considered. The listening tests are conducted on three sets of emotional utterances. Set-1 contains 40 original (5 *sentences* × 8 *emotions*) speech utterances chosen from a Telugu emotional database IITKGP-SESC [23]. The emotional speech files of set-1 represent the combined information of excitation source and vocal tract system. Set-2 represents only excitation source (LP residual) information of set-1 utterances. Set-3 contains the utterances representing only vocal tract information. The wave files of set-3 utterances are generated by exciting the VT parameters (LP Coefficients) of the utterances of set-1, using white random noise. Since the spectrum of white random noise is flat, the output of the VT system excited with random noise contains mainly the vocal tract characteristics. In each set, utterances are arranged randomly, before subjective listening tests are conducted. A group of 25 (16 males and 9 female) research students of the Indian Institute of Technology, Kharagpur, whose mother tongue is Telugu, are asked to recognize the emotions present in the utterances. The utterances are played to them, one after the other, and they are asked to categorize them in one of the given eight emotions. A gap of 5 min is provided before starting the evaluation of the next set. Each subject has to recognize the emotion category of 120 utterances (40 utterances from each set). The average emotion recognition performance of all three sets is given in Table 3.1, in the form of confusion matrices. The diagonal elements in each 8 × 8 matrix represent the percentage of correctly classified utterances. The average emotion recognition performance from the subjective listening tests is observed to be 47,

Table 3.1 Emotion classification performance using subjective listening tests. Emotional speech files are generated using (i) only excitation source signal, (ii) only vocal tract system features and (iii) both source and system features

Emotions	Emotion recognition performance (%)							
	Ang.	Dis.	Fear	Hap.	Neu.	Sad.	Sar.	Sur.
Subjective listening results using only source features (Set-1)								
Anger	40	17	7	10	10	0	6	10
Disgust	13	43	0	0	20	7	10	7
Fear	0	0	53	13	7	13	7	7
Happiness	10	0	17	37	17	6	0	13
Neutral	0	0	20	0	57	23	7	0
Sadness	3	10	20	7	10	50	0	0
Sarcasm	7	17	10	3	0	0	43	20
Surprise	3	10	10	7	0	0	20	50
Subjective listening results using only system features (Set-2)								
Anger	43	17	13	10	7	0	0	10
Disgust	20	47	0	0	17	10	6	7
Fear	0	0	63	10	7	10	0	10
Happiness	3	0	17	50	20	3	0	7
Neutral	3	0	13	10	57	10	7	0
Sadness	0	3	10	7	13	60	0	7
Sarcasm	0	10	7	3	0	0	57	23
Surprise	10	7	13	7	0	0	13	50
Subjective listening results using normal speech utterances (Set-3)								
Anger	70	13	3	3	7	0	0	4
Disgust	20	43	7	10	7	0	6	7
Fear	0	17	37	7	6	17	6	10
Happiness	3	3	7	67	13	0	0	7
Neutral	0	7	3	0	87	3	0	0
Sadness	0	13	20	0	13	54	0	0
Sarcasm	0	7	3	0	10	0	73	7
Surprise	0	3	23	17	10	0	0	47

Abbreviations: *Ang.* -Anger, *Dis.* -Disgust, *Hap.* -Happiness, *Neu.* -Neutral, *Sad.* -Sadness, *Sar.* -Sarcasm, *Sur.* -Surprise

53 and 60% using *source*, *system* and *source* + *system* respectively. The results indicate the presence of emotion-specific information in excitation source as well as vocal tract system.

To emphasize the intuition of presence of emotion-specific information in the excitation source, LP residual signals and GVV signals of the vowel segment /a/, taken from the speech utterance *thallidhandrulanu gauravincha valenu* of IITKGP-SESC are plotted for different emotions, in Figs. 3.1 and 3.2 respectively. It is observed that the strength of excitation around the glottal closure instants (GCI) are specific to each emotion. The instantaneous frequency (epoch intervals or pitch periods), phase changes at the GC regions and residual signal amplitude fluctuations are clear to be different for different emotions (see Fig. 3.1).

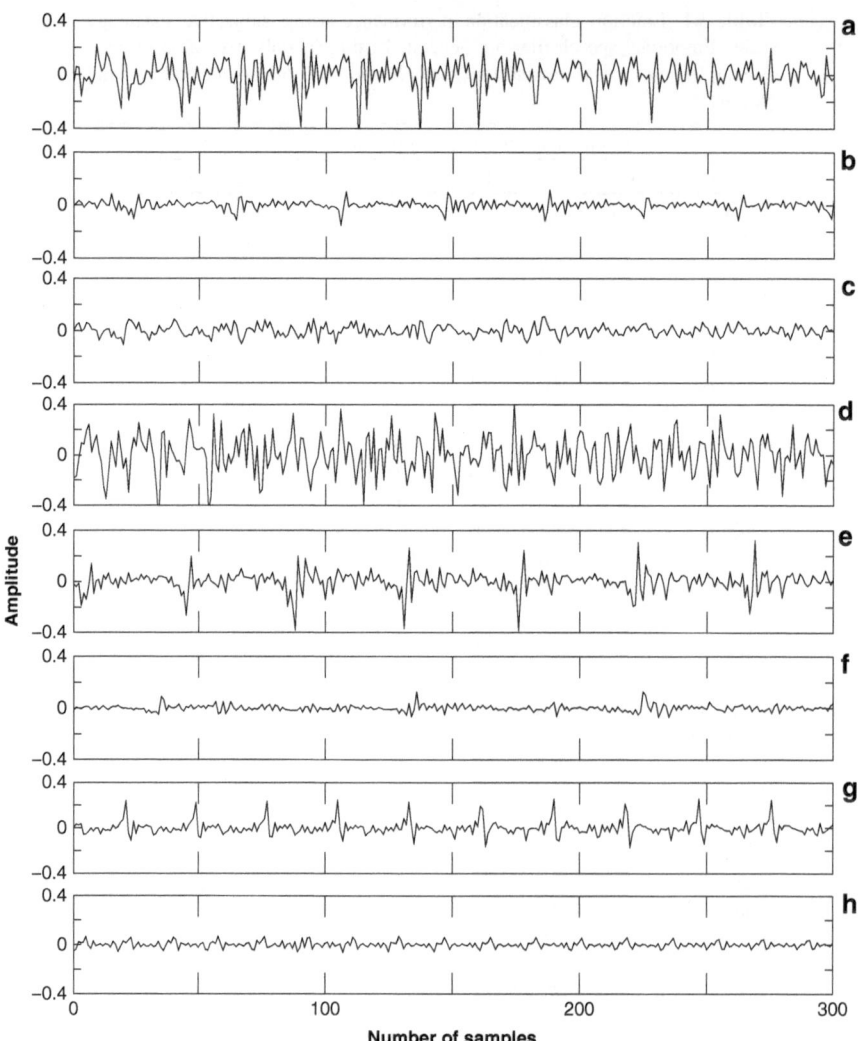

Fig. 3.1 LP residual signals for the vowel segment /a/, in eight emotions. (**a**) Anger, (**b**) Disgust, (**c**) Fear, (**d**) Happiness, (**e**) Neutral, (**f**) Sadness (**g**) Sarcasm, and (**h**) Surprises

From Fig. 3.2, it is to be noted that the variations in glottal pulse shapes (durations of glottal opening and glottal closure, their ratio) are unique to different emotions. The variations in amplitudes of GVV signal are to be noted here, as different scales are used to plot them. These measures, from the plots of LP residual and GVV signals, along with the results of subjective listening tests given in Table 3.2, have motivated us to explore the excitation source features for recognition of emotions from speech.

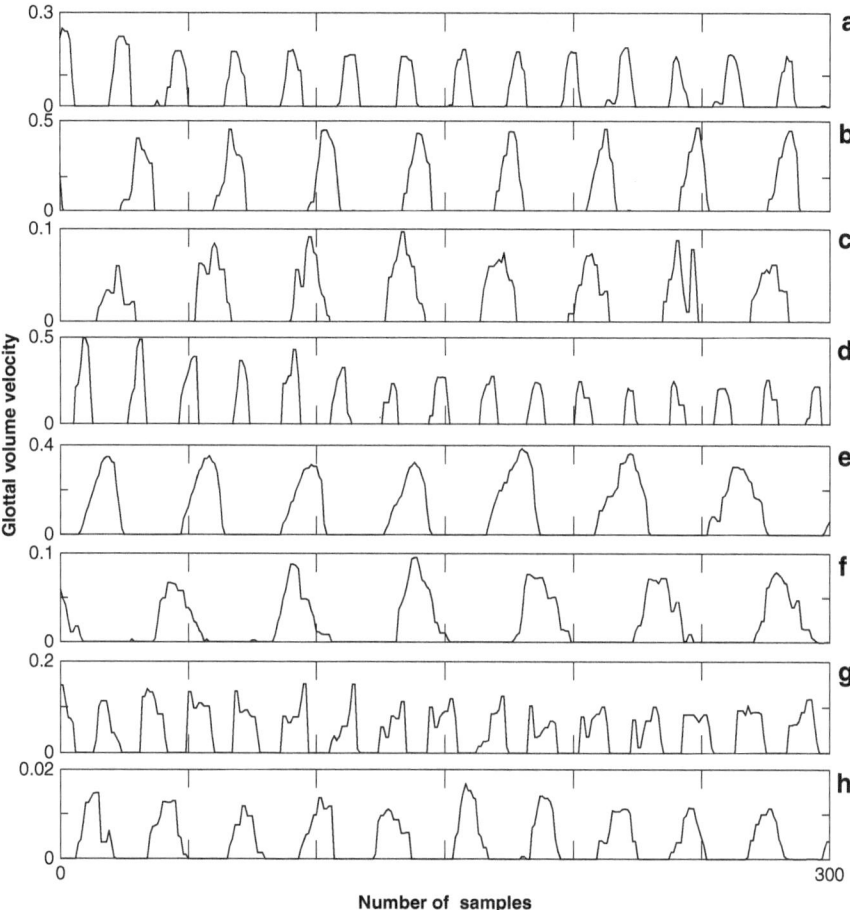

Fig. 3.2 Glottal volume velocity (GVV) signal for the vowel segment /a/, in eight emotions. (**a**) Anger, (**b**) Disgust, (**c**) Fear, (**d**) Happiness, (**e**) Neutral, (**f**) Sadness (**g**) Sarcasm, and (**h**) Surprise

3.3 Emotional Speech Corpora

In this work, the Indian Institute of Technology Kharagpur-Simulated Emotional Speech Corpus (IITKGP-SESC) and Berlin emotional speech database (Emo-DB) are used to analyze the performance of proposed excitation source features for Emotion Recognition (ER).

3.3.1   Indian Institute of Technology Kharagpur-Simulated Emotional Speech Corpus: IITKGP-SESC

This database is particularly designed and developed at the Indian Institute of Technology, Kharagpur, to support the study on speech emotion recognition. The proposed speech database is the first one developed in an Indian language (Telugu), for analyzing the common emotions present in day-to-day conversations. This corpus is sufficiently large to analyze the emotions in view of speaker, gender, text and session variability.

The corpus is recorded using ten (five male and five female) professional artists from All India Radio (AIR) Vijayawada, India. The artists are sufficiently experienced in expressing the desired emotions from the neutral sentences. All the artists are in the age group of 25–40 years, and have the professional experience of 8–12 years. For analyzing the emotions we have considered 15 semantically neutral Telugu sentences. Each of the artists has to speak the 15 sentences in 8 given emotions in one session. The number of sessions considered for preparing the database is 10. The total number of utterances in the database is 12,000 (15 *sentences* × 8 *emotions* × 10 *artists* × 10 *sessions*). Each emotion has 1,500 utterances. The numbers of words and syllables in the sentences are varying from 3–6 and 11–18 respectively. The total duration of the database is around 7 h. The eight emotions considered for collecting the proposed speech corpus are: Anger, Disgust, Fear, Happiness, Neutral, Sadness, Sarcasm and Surprise. The speech samples are recorded using SHURE dynamic cardioid microphone C660N. The distance between the microphone and the speaker is maintained approximately around 3 and 4 in. The speech signal is sampled at 16 kHz, and each sample is represented as a 16-bit number. The sessions are recorded on alternate days to capture the inevitable variability in the human vocal tract system. In each session, all the artists have given the recordings of 15 sentences in 8 emotions. The recording is done in such a way that each artist has to speak all the sentences at a stretch in a particular emotion. This provides the coherence among the sentences for each emotion category. The entire speech database is recorded using a single microphone and at the same location. The recording was done in a quiet room, without any obstacles in the recording path.

The quality of the database is also evaluated using subjective listening tests. Here, the quality represents how well the artists simulated the emotions from the neutral text. The subjects are used to assess the naturalness of the emotions embedded in speech utterances. This evaluation is carried out by 25 post graduation and research students of the Indian Institute of Technology, Kharagpur. This subjective listening test is useful for the comparative analysis of emotions in a human versus machine perspective. In this study, 40 sentences (5 sentences from each emotion) randomly selected from male and female speakers are considered for evaluation. Before taking the test, the subjects are given the pilot training by playing eight sentences (a sentence from each emotion) from each artist's speech data, for understanding

Table 3.2 Emotion classification performance based on subjective evaluation

Emotions	Emotion recognition performance															
	Male artist								Female artist							
	A	D	F	H	N	Sa	S	Sur	A	D	F	H	N	Sa	S	Sur
Anger	73	17	2	3	4	0	0	1	69	19	3	2	5	0	0	2
Disgust	28	56	7	0	4	0	5	0	40	44	5	0	3	2	3	3
Fear	7	6	49	0	8	19	1	10	6	8	37	2	7	25	1	14
Happiness	0	2	6	62	8	9	5	6	0	4	4	66	10	7	3	6
Neutral	0	5	0	6	86	2	0	1	1	8	1	6	83	0	1	0
Sadness	0	3	16	3	13	61	4	0	4	2	25	1	12	52	3	1
Sarcasm	6	5	0	0	0	4	85	0	4	5	0	0	0	3	88	0
Surprise	0	7	5	16	5	5	7	55	6	6	3	17	3	16	1	48

A Anger, *D* Disgust, *F* Fear, *H* Happiness, *N* Neutral, *Sa* Sadness, *S* Sarcasm, *Sur* Surprise

(familiarizing) the characteristics of emotion expression. Forty sentences used in this evaluation are randomly ordered, and played to the listeners. For each sentence, the listener has to mark the emotion category from the set of eight given emotions. The overall emotion classification performance for male and female speech data is given in Table 3.2. The observation shows that the average emotion recognition rates of male and female speech utterances are 61 and 66% respectively.

In this book, emotion recognition studies have been carried out in three ways, accordingly three data sets are derived from IITKGP-SESC. They are (1) Set1 – Session independent speech emotion recognition, (2) Set2 – Text independent speech emotion recognition, and (3) Set3 – Speaker and text independent speech emotion recognition. Set1 is used to analyze the emotion recognition in view of session variability. Here eight sessions of all speakers' speech data is used for training the emotion models, and the remaining two sessions of all speakers' speech data is used for testing. Set2 is used to study emotion recognition in view of text independent speech data. Here, eight sessions of all speakers' speech data containing the first ten text prompts are used for training, and while testing the remaining five text prompts of the last two sessions from all the speakers' speech data are used. Set3 is used to analyze the emotion recognition with respect to speaker and text independent speech data. Here, training is performed with eight speakers' (four males and four females) speech data, from all ten sessions. Testing is performed with the remaining two speakers' (one male and one female) speech data. To realize the text independent data, during training the speech utterances corresponding to the first ten text prompts of the database are used, and the remaining five text prompts are used while testing. Set3 is designed to have text and speaker independent properties and is more generalized than the other two sets. Therefore the majority of the results discussed in this 20 are derived using the Set3 dataset. Brief results of Set1 and Set2 are discussed at the end of the respective chapters. Table 3.3 shows the details of the three datasets used in this work.

Table 3.3 Details of the datasets derived from IITKGP-SESC, for various studies on speech emotion recognition

Data set	Purpose and description	Training data	Testing data
Set1	Session independent emotion recognition	The utterances of all 15 text prompts, recorded from 10 speakers are used in training. Out of 10 sessions, 8 (1–8) sessions of each speaker are used in training	The utterances of all 15 text prompts, recorded from 10 speakers are used in testing. Out of 10 sessions, 2 (9 and 10) sessions of each speaker are used in testing
Set2	Session and text independent emotion recognition	Out of 15 text prompts, the utterances of 10 (1–10) prompts, recorded from 10 speakers are used in training. Out of 10 sessions, 8 (1–8) sessions of each speaker are used in training	Out of 15 text prompts, the utterances of 5 (11–15) prompts, recorded from 10 speakers are used in testing. Out of 10 sessions, 2 (9 and 10) sessions of each speaker are used in testing
Set3	Session, text, and speaker independent emotion recognition	Out of 15 text prompts, the utterances of 10 (1–10) prompts, recorded from 8 (4 males and 4 females) speakers are used in training. All ten sessions of each speaker are used in training	Out of 15 text prompts, the utterances of 5 (11–15) prompts, recorded from 2 (a male and a female) speakers are used in testing. All ten sessions of each speaker are used in testing

3.3.2   Berlin Emotional Speech Database: Emo-DB

F. Burkhardt et al. have collected the actor-based simulated-emotion Berlin database in the German language [22]. Ten (five male + five female) actors have contributed in preparing the database. The emotions recorded in the database are anger, boredom, disgust, fear, happiness, neutral and sadness. Ten linguistically neutral German sentences are chosen for database construction. The database is recorded using the Sennheiser MKH 40 P48 microphone, with the sampling frequency of 16 kHz. Samples are stored as 16 bit numbers. Eight hundred and forty (840) utterances of Emo-DB are used in this work.

In the case of the Berlin database, eight speakers' speech data is used for training the models and the remaining two speakers' speech data is used for validating the trained models.

3.4 Excitation Source Features for Emotion Recognition

In this work, the LP residual represents excitation source information. The LP residual signal is derived from a speech signal using tenth order LP analysis. Among different categories of excitation, voiced excitation contains significant emotion-specific information, because the corresponding glottal vibration pattern may be different for different emotions. Rate of glottis vibration, variation in GVV signals, strength and sharpness of significant excitation at glottal closure attribute to the characteristics of specific emotion. In this work, the higher order relations among the LP residual signal, LP residual phase, epoch parameters, GVV signal parameters are proposed as excitation source features for recognizing the emotion from speech. The extraction of proposed excitation source features is discussed in the following subsections.

3.4.1 Higher-Order Relations Among LP Residual Samples

In the linear prediction analysis of speech, each sample is predicted as a linear weighted sum of the past p samples, where p represents the order of prediction [76]. If $s(n)$ is the present sample, then it is predicted by the past p samples as

$$\hat{s}(n) = -\sum_{k=1}^{p} a_k s(n-k)$$

The difference between the actual and predicted sample value is termed as the prediction error or residual, which is given by

$$e(n) = s(n) - \hat{s}(n) = s(n) + \sum_{k=1}^{p} a_k s(n-k)$$

where a_k's are the linear prediction coefficients. The linear prediction coefficients are typically determined by minimizing the mean squared error over an analysis frame. The coefficients can be obtained by solving the set of p normal equations using the autocorrelation method,

$$\sum_{k=1}^{p} a_k R(n-k) = -R(n), \qquad n = 1, 2, \ldots, p$$

where

$$R(k) = \sum_{n=0}^{N-(p-1)} s(n)s(n-k), \qquad k = 0, 1, 2, \ldots, n$$

$R(k)$ is autocorrelation function. The residual signal $(e(n))$, shown above is obtained by passing the speech signal through the inverse filter $A(z)$, given by

$$A(z) = 1 + \sum_{k=1}^{p} a_k z^{-k}$$

The LP spectrum $|H(w)|^2$ is given by

$$|H(w)|^2 = \left| \frac{G}{1 + \sum_{k=1}^{p} a_k e^{-jwk}} \right|^2$$

where G is the gain parameter given by the minimum mean squared error.

An LP order of 8–14 seems to be appropriate for deriving the LP residual, for a speech signal sampled at 8 kHz. In this work we use the speech signal sampled at 8 kHz and the LP order 10 for deriving the LP residual. The LP residual signal mostly contains the excitation source information. More details about LP analysis are given in Appendix A.

Since LP analysis extracts out the first and second order relations present among the adjacent samples through autocorrelation coefficients, the LP residual does not contain any significant lower order relations corresponding to shape and size of the vocal tract. This is the reason that the autocorrelation function of the LP residual signal has low correlation values for nonzero time lags [14]. Therefore we are exploring higher-order relations among the samples of the LP residual signal for investigating the presence of emotion-specific information. Capturing the emotion-specific information from these higher-order relations may involve a nonlinear process. Since neural network (NN) models can be configured to capture the nonlinear information present among the samples, we use these models in this study. In particular, auto-associative neural network (AANN) models are explored to extract the desired emotion-specific information from residual samples. To demonstrate the presence of higher order relations among the samples of an LP residual, training errors of AANN are plotted by training the network with the samples of LP residual and noise. It is observed from Fig. 3.3 that the training error using LP residual samples is decreasing exponentially as the number of training iterations is increased. This indicates that LP residual signal contains some higher order correlations among the samples. The training error using samples of noise is observed to be non-decreasing, indicating no relations among the noise samples.

3.4.2   Phase of LP Residual Signal

The rapid fluctuations in the amplitudes of LP residual is mainly due to fluctuation in the phase polarities. Speaker recognition studies have demonstrated that the phase of an LP residual signal contains the speaker specific information [129]. In this work characteristics of LP residual phase are explored for extracting the emotion-specific information present in the excitation source signal.

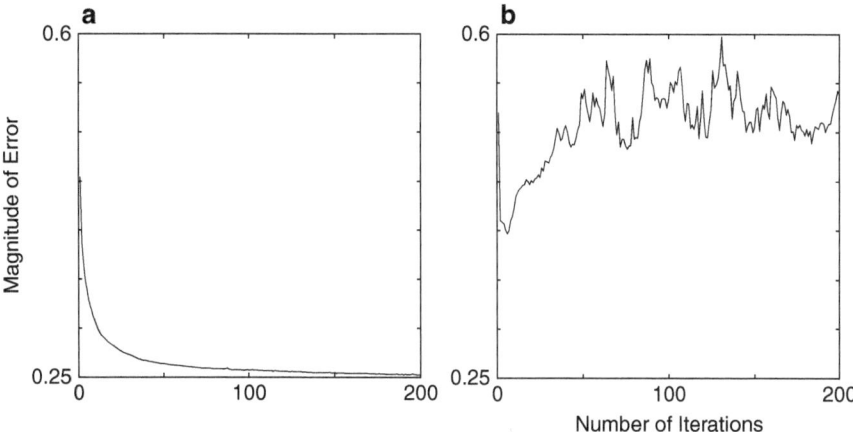

Fig. 3.3 Training error patterns of AANN models. (**a**) with LP residual and (**b**) with random noise

The residual phase is derived as the cosine of the phase function of the analytic signal $r_a(n)$ derived from a LP residual of a speech signal $r(n)$. The analytic signal $r_a(n)$ corresponding to $r(n)$ is given by

$$r_a(n) = r(n) + jr_h(n)$$

where $r_h(n)$ is the Hilbert transform of $r(n)$, and is given by

$$r_h(n) = IFT[r_h(\omega)]$$

where

$$r_h(\omega) = \begin{cases} -jR(\omega), \ 0 \le \omega < \pi \\ jR(\omega), \quad 0 > \omega \ge -\pi \end{cases}$$

Here $R(\omega)$ is the Fourier transform of $r(n)$ and IFT denotes inverse Fourier transform. The magnitude of analytic signal $r_a(n)$ (Hilbert envelope $h_e(n)$) is given by

$$h_e(n) = |r_a(n)| = \sqrt{r^2(n) + r_h^2(n)}$$

and the cosine of the phase of the analytic signal $r_a(n)$ is given by

$$\cos(\Theta(n)) = \frac{Re(r_{a(n)})}{r_a(n)} = \frac{r(n)}{h_e(n)}$$

A segment of voiced speech, its LP residual, Hilbert envelope of LP residual and phase of the LP residual are shown in Fig. 3.4. It is difficult to observe any perceivable relationships among the samples of LP residual phase from the plot,

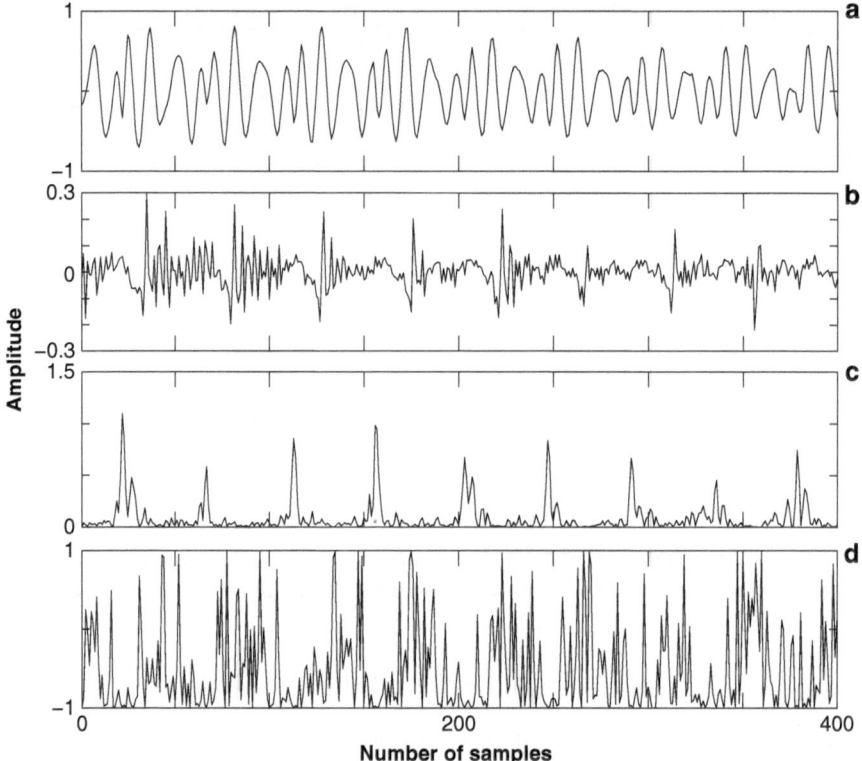

Fig. 3.4 (a) Segment of a voiced speech signal, (b) LP residual of a speech signal shown in (a), (c) Hilbert envelope of LP residual and (d) Phase of the LP residual signal

however, since during LP analysis only the second order relations are removed, the higher order relations among the samples of the speech signal are reflected in the samples of residual phase. AANN models are used to capture these higher order relations specific to each emotion.

3.4.3   Parameters of the Instants of Glottal Closure (Epoch Parameters)

In speech, the instants of significant excitation correspond to the instants of glottal closure, in the case of voiced speech and to some random excitation like the onset of burst, in the case of non-voiced speech. These instants of significant excitation are also known as epochs. In this book, the terms GCI, ISE, and epochs are used interchangeably. In a speech signal, the region around glottal closure instants (GCI), within a pitch period corresponds to high SNR due to impulse-like excitation. At the instants of glottal closure the prediction is poor during the LP analysis and results into larger error. During the glottal closure interval, the vocal tract system (oral cavity) is completely isolated from the trachea. In the glottal open phase the vocal tract (oral/nasal cavity) is coupled with trachea. The estimate of excitation source signal is more accurate during the glottal closure interval, due to complete isolation of vocal tract from trachea. Therefore, we are exploring the properties of excitation source signal in the vicinity of instants of glottal closure to capture the emotion-specific information. In this work, the characteristics of instants of significant excitation or epoch is represented by four parameters, namely strength of epoch, epoch intervals (instantaneous frequency), sharpness of epoch and slope of strengths of epochs.

Detection of epoch locations: Locating the instants of significant excitation (epoch) is necessary for determining the epoch parameters. A zero frequency filter-based approach is used to determine the locations of epochs and instantaneous frequency. Instantaneous frequency is a pitch value computed at epoch locations. It is estimated as the reciprocal of the time period between two consecutive epoch locations. The impulse-like excitation is due to abruptness of the glottal closure in each cycle. The characteristics of the sequence of impulse-like excitations are reflected across all the frequencies in the speech signal including 0 Hz. Filtering the speech signal through a resonator located at zero frequency helps in emphasizing the characteristics of excitation [130]. The system function of such a resonator is given by

$$H(z) = \frac{1}{1 + a_1 z^{-1} + a_2 z^{-2}}$$

where $a_1 = -2$ and $a_2 = 1$. The above resonator de-emphasizes the characteristics of the vocal tract, since the resonances of the vocal tract are located at much higher frequencies than the zero frequency. A cascade of two such resonators, given by the system function

$$G(z) = H(z)H(z)$$

is used to reduce the effect of all the resonances of vocal tract. Let $s[n]$ denote the input speech signal. Then the output $x_s[n]$ of the cascade of the resonator is given by

$$x_s[n] = s[n] \star g[n]$$

where $g[n]$ is the impulse response of the system function $G(z)$ and '$\star$' denotes the convolution operation. The output $x_s[n]$ contains mainly the features of glottal vibrations. Filtering the signal $s[n]$ through the cascade of resonators causes the output $x_s[n]$ to grow as a polynomial function of time. This trend in $x_s[n]$ is removed by subtracting the average of sample values over a window of 10 ms. (approximately 0.5–1.5 times the estimated pitch period). The resulting trend removed signal $y_s[n]$ is given by

$$y_s[n] = x_s[n] - \frac{1}{2N+1} \sum_{k=-N}^{N} x_s[n+k]$$

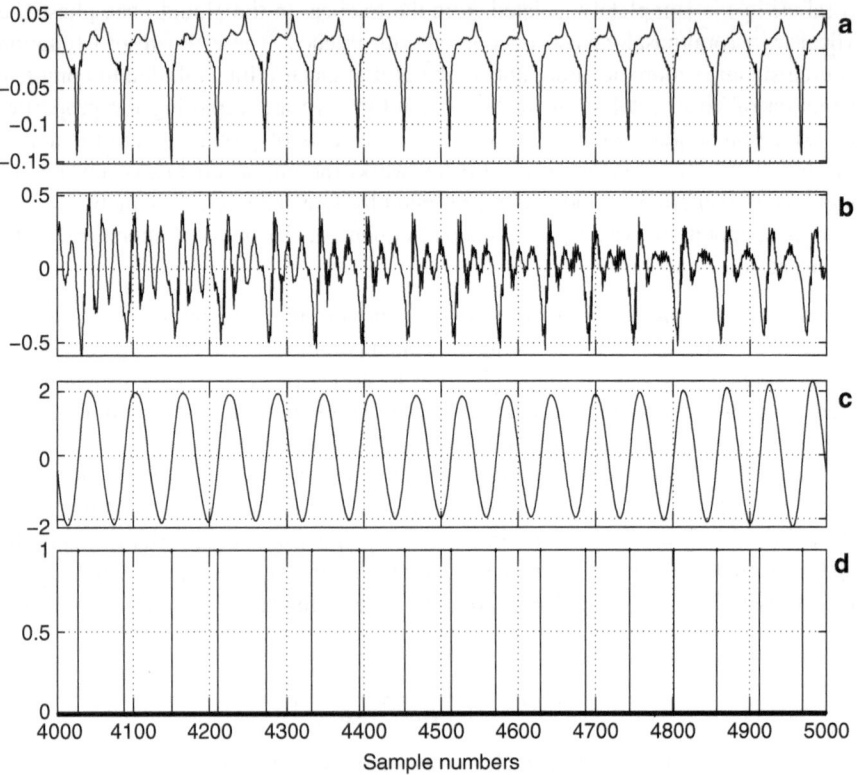

Fig. 3.5 (**a**) EGG signal (reference epoch locations), (**b**) segment of a speech signal, (**c**) zero frequency (narrow band pass) filtered speech signal and (**d**) epoch locations

where $2N + 1$ is the size (in samples) of the window [130]. The signal $y_s[n]$ is called the filtered signal. Figure 3.5c shows the zero frequency (narrow band pass) filtered signal for the segment of the voiced speech shown in Fig. 3.5b. Positive zero crossings in zero frequency filtered speech signal are approximated as the epoch locations. Figure 3.5d shows the approximated epoch locations. To validate the results obtained, the differenced electro-glottolo-graph (EGG) signal is shown in Fig. 3.5a. The vibration pattern of glottal folds can be directly recorded from the electro-glottolo-graph. It is a device, that can be attached to the throat of a speaker, and a transducer in it converts pressure variations into an electrical signal. From Fig. 3.5a, d, it may be observed that the approximated and actual epoch locations are almost matching.

The strength of excitation can be viewed as the amplitude of the impulse like excitation at the epoch location. Determining the strength of excitation from the LP residual signal is difficult due to the fluctuation in residual sample amplitudes within the GC region due to phase fluctuation. The effect of impulse like excitation

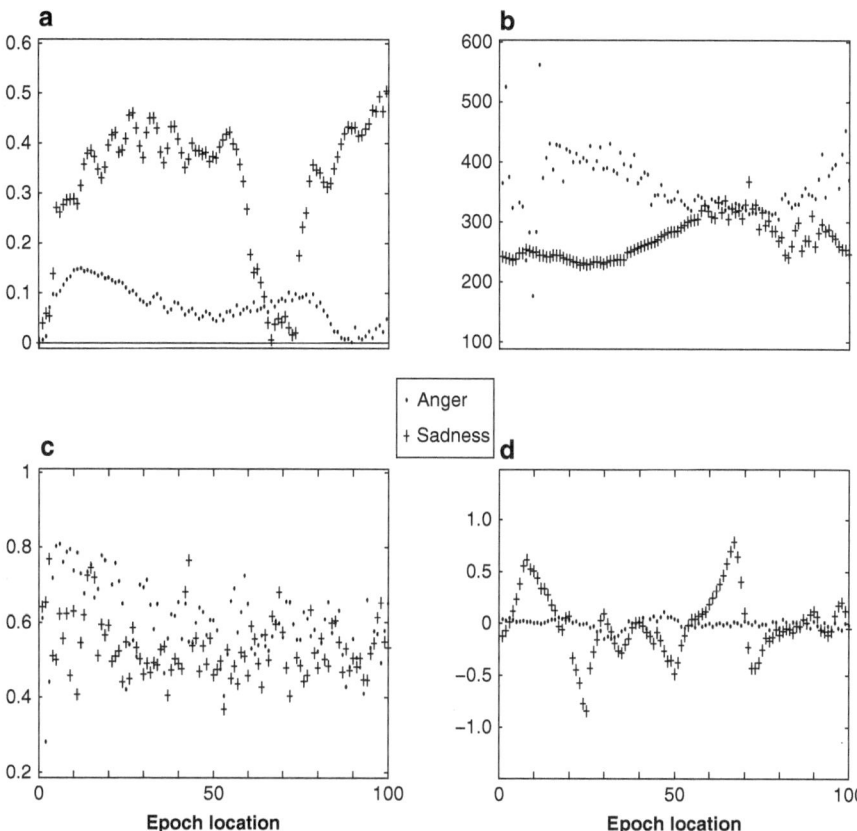

Fig. 3.6 Epoch parameter values for the sequence of 100 epochs, for the utterances of anger and sadness. (**a**) Strength of epochs, (**b**) Instantaneous frequency, (**c**) Sharpness of epochs and (**d**) Slope of strength of epochs

is spread uniformly across the frequency range, hence relative strength of excitation can be derived from a narrow band around any frequency. As the vocal tract system has resonances at much higher frequencies than zero frequency, the information about the strength of excitation is derived from the zero frequency resonated speech signal. The slope of the zero frequency filtered speech signal around the zero crossings, corresponding to the epoch locations, gives a measure of strength of excitation. Figure 3.6a shows the strength of excitation values for anger and sadness emotions.

Sharpness of epoch can be calculated from the Hilbert envelope of the LP residual signal using the formulation $\eta = \sigma/\mu$. It is computed as the ratio of standard deviation (SD) and mean of residual samples within a 3 ms window centered at each epoch location. For a segment of length N samples consisting of an ideal impulse (in the discrete-time domain) of amplitude a, $\eta = \sqrt{N}$. For a segment of length N consisting of samples of equal amplitude $a/\sqrt{N}$, $\eta = 0$. The segment in this case has the same energy as that of the ideal impulse of amplitude a. This case represents the maximum deviation from an ideal impulse. Thus the value of η lies between 0 and $\sqrt{N}$ for any segment, irrespective of the amplitudes of the samples in the segment. A higher value of η indicates greater sharpness of excitation. In general, a segment having impulse like characteristics in excitation, as in the case of anger emotion, has a smaller value of μ and a larger value of σ, resulting in a larger value of η. By contrast, a sadness emotion utterance with greater spread around the center has a larger value of μ and a smaller value of σ, resulting in a smaller value of η [40].

Slope of strength of excitation at each epoch can be computed by considering the strengths of epochs within a window of five consecutive pitch periods. The deviation between the values of the first and the fifth epoch strengths gives the slope. The window is shifted by one epoch interval for determining the next value of slope of strength of epoch. Window is shifted by one epoch interval for determining the next value of slope of strength of epochs. In this manner the slope of strength of epoch is computed, by shifting the window by an epoch interval each time. The values of 4 epoch parameters for the sequence of 100 epochs are plotted in Fig. 3.6. The emotions considered are anger and sadness. This figure illustrates the discriminative properties of epoch parameters. Only two emotions are plotted to avoid confusion due to overlapping of data points. Similarly, Fig. 3.7 shows the histograms of the distribution of four epoch parameters for all eight emotions considered in this study. Histogram for each of the epoch parameters is plotted using appropriate number of bins to visualize proper distribution. A column represents the histograms of specific epoch parameter for all eight emotions. Considering a specific epoch parameter, density of the points in different bins and the region of their spread are perceivably distinct for different emotions. The partial overlapping of distribution of points may lead to misclassification of emotions, while categorizing them.

3.4.4   Dynamics of Epoch Parameters at Syllable Level

To capture the emotion-specific knowledge present in the sequence of epochs at the syllable level, the epoch parameters of all epochs present in the syllable are concatenated. This leads to the feature vectors of different sizes based on the number of epochs present in the syllable. For attaining the fixed dimensional feature vectors, 20 epochs are considered per syllable (20 *epochs* × 4 *parameters per epoch* = 80 *parameters*). If the number of epochs in the syllable is more than 20, then extra epochs are omitted at uniform intervals or if the number of epochs is less than 20, then some of the epochs are replicated from the steady vowel portion of that syllable. AANN models are developed for each emotion to capture the dynamics of epoch parameters at the syllable level.

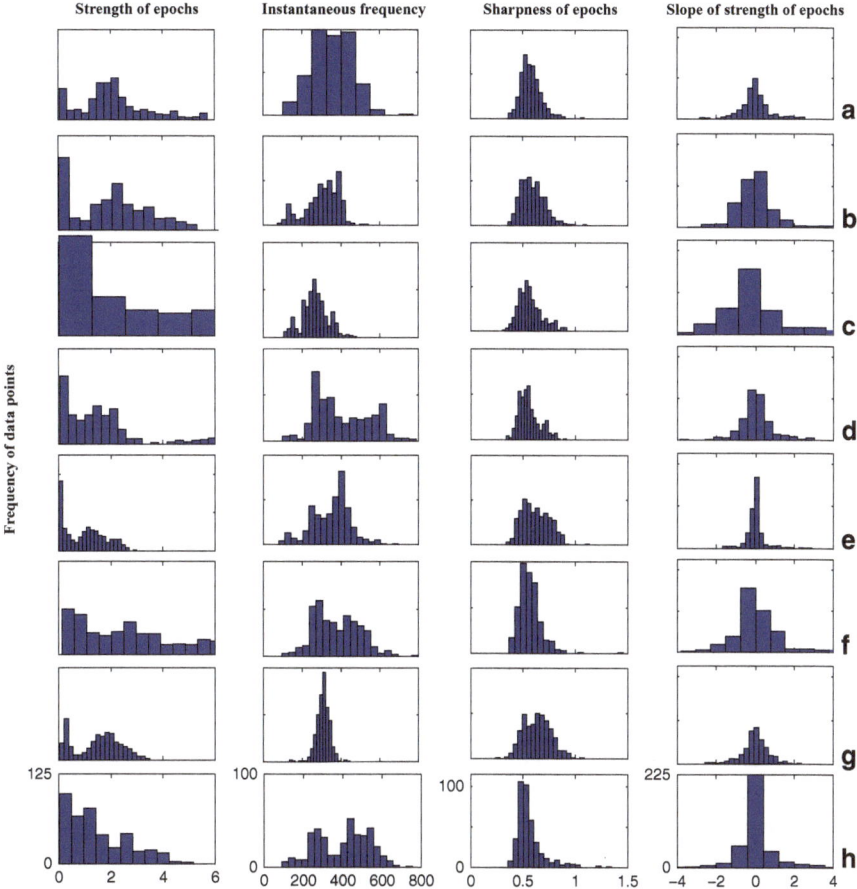

Fig. 3.7 Histograms of the epoch parameters for (**a**) Anger, (**b**) Disgust, (**c**) Fear, (**d**) Happiness, (**e**) Neutral, (**f**) Sadness, (**g**) Sarcasm, and (**h**) Surprise

3.4.5   Dynamics of Epoch Parameters at Utterance Level

While analyzing emotional conversation, it is observed that the sequence of words in an utterance undergoes various emotion-specific modulations through varying intonation and energy patterns while expressing different emotions. To capture the emotion-specific information present in the sequence of words of an utterance, epoch parameters are extracted from an utterance are concatenated to derive the feature vector. In this work, 50 epochs are chosen at uniform interval from an entire sequence of epochs to represent the dynamics of epoch parameters of the utterance. Therefore the feature vectors of dimension 200 (50 *epochs* × 4 *parameters per epoch*) are generated for each utterance. Support vector machines

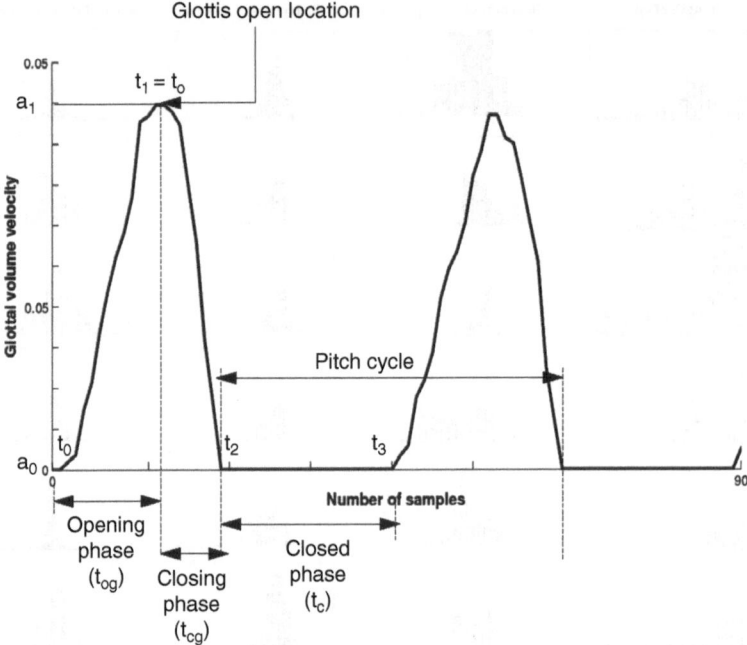

Fig. 3.8 Two cycles of glottal volume velocity signal

are used for discriminating the emotions using the utterance level epoch parameters. Here SVMs are preferred mainly because of availability of few feature vectors in high dimensional feature space.

3.4.6   Glottal Pulse Parameters

The glottal pulse signal also known as the glottal volume velocity (GVV) signal is obtained by passing the LP residual signal through a low pass filter. In the discrete domain low pass filtering can be implemented by an integration operation. It is observed from Fig. 3.2 that GVV parameters like (a) duration of opening phase, (b) duration of closing phase, (c) duration of complete closure, (d) slope of the opening phase, (e) slope of the closing phase, (f) ratio of the slope of open phase to the slope of closing phase, have enough discrimination among different emotions. Figure 3.8 shows two cycles of GVV signal and its parameters. Six GVV parameters mentioned above are computed for each glottal cycle. In this work, along with the GVV parameters, the higher order relations among the samples of GVV signal are also explored to capture the emotion-specific information. Boundaries of the glottal cycle are determined using epoch locations. The higher order relations among the samples of GVV signal in each pitch period are captured in a similar way as the samples of the LP residual signal are captured.

3.5 Classification Models

In this work AANN and SVM models are used to capture the emotion-specific information from excitation source features. AANN models are used to capture the nonlinear relations among the feature vectors. SVMs perform the classification based on the discriminative information present in the feature vectors.

3.5.1 Auto-associative Neural Networks

AANN models are basically feed-forward neural network (FFNN) models, which try to map an input vector onto itself, and hence the name auto-association or identity mapping [131, 132]. It consists of an input layer, an output layer and one or more hidden layers. The number of units in the input and output layers is equal to the dimension of the input feature vectors. The number of nodes in one of the hidden layers is less than the number of units in either the input or output layer. This hidden layer is also known as the dimension compression layer. The activation function of the units in the input and output layers is linear, whereas in the case of hidden layers it is either linear or nonlinear. A five-layer AANN model with the structure shown in Fig. 3.9 is used in this study. The structure of network used to capture the higher order relations, from the sequence of LP residual samples is 40L-60N-12N-60N-40L, where L refers to linear units and N to nonlinear units. The integer value indicates the number of units present in that layer. The number of linear elements at the input layer indicates, the size of the feature vectors used for developing the

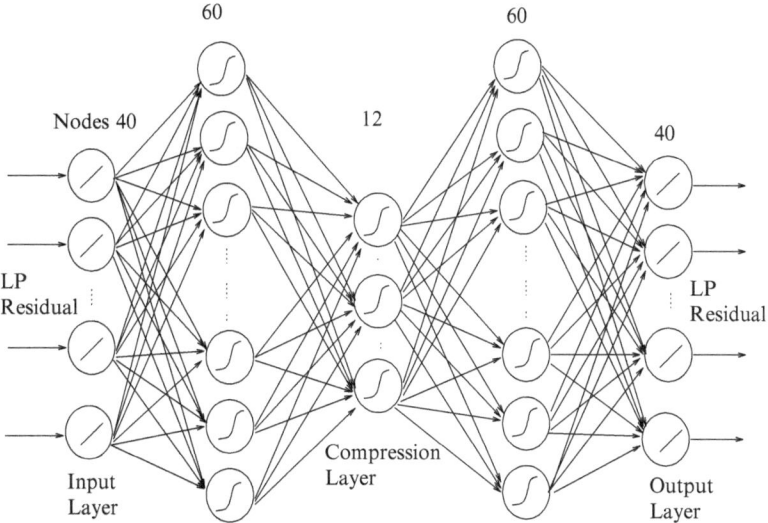

Fig. 3.9 Five layer auto-associative neural network

models. The nonlinear units use $tanh(s)$ as the activation function, where s is the net input value of that unit. The structure of the network was determined empirically.

The performance of AANN models can be interpreted in different ways, depending on the application and the input data [133, 134]. If the data is a set of feature vectors in the feature space, then the performance of AANN models can be interpreted as linear or nonlinear principal component analysis (PCA) or capturing the distribution of input data [135–137]. On the other hand, if the AANN is presented directly with signal samples, such as LP residual signal, the network captures the implicit linear/ nonlinear relations among the samples [138–140].

Determining the network structure is an optimization problem. At present there are no formal methods for determining the optimal structure of a neural network. The key factors that influence the neural network structure are learning ability of a network and capacity to generalize the acquired knowledge. From the available literature, it is observed that five layer symmetric neural networks, with three hidden layers have been used for different speech tasks. The first and the third hidden layers have more number of nodes than the input or output layer. The middle layer (also known as dimension compression layer) contains fewer units [141, 142]. In this type of network, generally the first and third hidden layers are expected to capture the local information among the feature vectors and the middle hidden layer is meant for capturing global information. Most of the existing studies [141–145] have used the five layer AANNs with the structure $N_1L - N_2N - N_3N - N_2N - N_1L$, for their optimal performance. Here N_1, N_2, and N_3 indicate the number of units in the first, second and third layers respectively, of the symmetric five-layer AANN. Usually N_2 and N_3 are derived experimentally, for achieving the best performance in the given task. From the existing studies, it is observed that N_2 is in the range of 1.3–2 times N_1 and N_3 is in the range of 0.2–0.6 times N_1. For designing the structure of the network, we have used the guidelines from the existing studies and experimented with few structures for finalizing the optimal structure. The final network structure of 40L-60N-12N-60N-40L is chosen for exploring the emotion-specific information from the samples of LP residual and GVV signals. The performance of the network does not depend critically on the structure of the network [139, 146–148]. The number of units in the two hidden layers is guided by the heuristic arguments given above. All the input and output features are normalized to the range $[-1, +1]$ before presenting to the neural network. The back-propagation learning algorithm is used for adjusting the weights of the network to minimize the mean squared error [142].

3.5.2 Support Vector Machines

A notable characteristic of a support vector machine (SVM) is that the computational complexity is independent of the dimensionality of the kernel space, where the input feature space is mapped. Thus the curse of dimensionality is bypassed. SVMs have been applied to a number of different applications ranging from handwritten digit recognition to person identification and medical imaging.

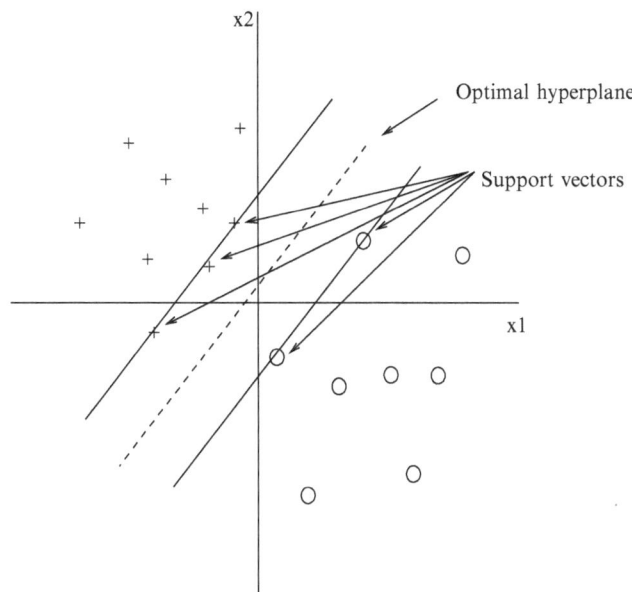

Fig. 3.10 Classification mechanism of support vector machine

The results shown in these studies indicate that, SVM classifiers exhibit enhanced generalization performance [149]. However, intelligent design or choice of kernel function adds to the real strength of support vector machines. SVMs are designed for two-class pattern classification. Multiclass (n-class) pattern classification problems can be solved using a combination of binary (2-class) support vector machines. A one-against-the-rest approach is used for decomposition of an n-class pattern classification problem into n two-class classification problems. The set of training examples $\left\{\{(x_i,k)\}_{i=1}^{N_k}\right\}_{k=1}^{n}$ consists of N_k number of examples belonging to the kth class, where the class label $k \in \{1,2,3,\ldots,n\}$. All the training examples are used to construct the SVM for a class. The SVM for the class k is constructed using the set of training examples and their desired outputs, $\left\{\{(x_i,y_i)\}_{i=1}^{N_k}\right\}_{k=1}^{n}$ The desired output y_i for the training example x_i is defined as follows:

$$y_i = \begin{cases} +1 \text{ if } x_i \in k\text{th class} \\ -1 \text{ otherwise} \end{cases}$$

The examples with $y_i = +1$ are called positive examples, and those with $y_i = -1$ are negative ones. An optimal hyperplane is constructed to separate positive examples from negative ones. The separating hyperplane (margin) is chosen in such a way as to maximize its distance from the closest training examples of different classes. Figure 3.10 illustrates the geometric construction of the hyperplane for a two-dimensional input space. The support vectors are those data points that lie

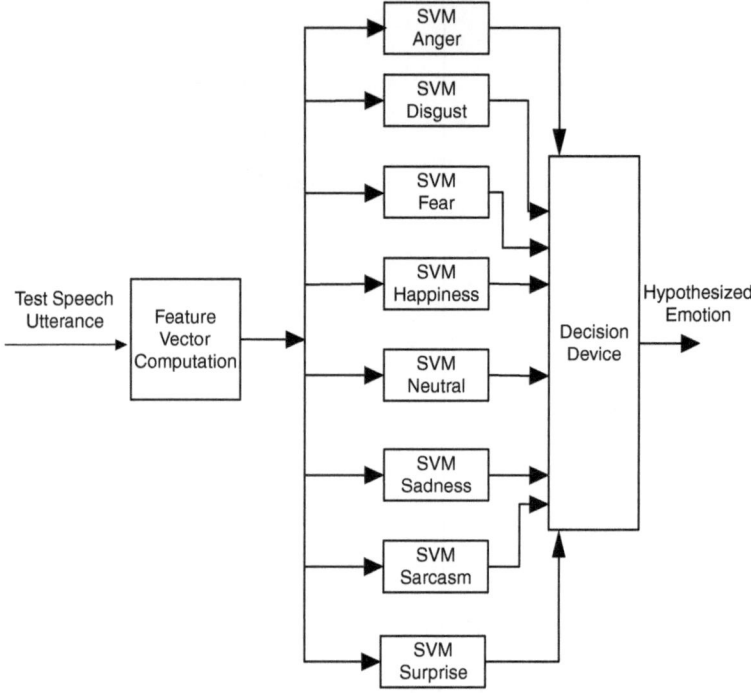

Fig. 3.11 General block architecture of an emotion recognition system using SVMs

closest to the decision surface, and therefore the most difficult to classify. They have a direct bearing on the optimum location of the decision surface [150]. For a given test pattern x, the evidence $D_k(x)$ is obtained from each of the SVMs. In the decision logic, the class label k associated with SVM, which gives maximum evidence, is hypothesized as the class (C) of the test pattern, that is

$$C(x) = argmax(D_k(x))$$

The block diagram of the emotion recognition system, developed using SVMs, is show in Fig. 3.11.

3.6 Results and Discussion

In this work, the Berlin emotional speech corpus and IITKGP-SESC are used to analyze excitation source features for speech emotion recognition. In the case of the Berlin emotional speech corpus, emotional speech utterances of eight speakers are used for developing emotion models, and remaining two speakers' speech data is used for validating the trained models. In the case of the IITKGP-SESC

Table 3.4 Emotion classification performance using samples of LP residual. Model: ERS-1(a), Average emotion recognition performance: 46.38

Emotions	Emotion recognition performance (%)							
	Anger	Disgust	Fear	Happiness	Neutral	Sadness	Sarcasm	Surprise
Anger	47	0	17	13	23	0	0	0
Disgust	0	57	0	0	13	17	0	10
Fear	13	20	37	0	3	10	0	17
Happiness	0	23	17	40	7	0	0	13
Neutral	0	10	0	0	60	23	7	0
Sadness	3	7	7	0	23	50	0	10
Sarcasm	3	27	17	10	0	0	43	0
Surprise	0	0	27	23	3	10	0	37

Table 3.5 Emotion classification performance using LP residual phase. Model: ERS-2(a) Average emotion recognition performance: 41.38

Emotions	Emotion recognition performance (%)							
	Anger	Disgust	Fear	Happiness	Neutral	Sadness	Sarcasm	Surprise
Anger	37	0	17	23	13	0	0	10
Disgust	0	40	0	7	23	23	0	7
Fear	20	17	30	10	13	7	0	3
Happiness	13	20	0	43	17	7	0	0
Neutral	0	17	0	0	47	23	13	0
Sadness	0	23	0	0	20	50	7	0
Sarcasm	0	17	23	0	0	0	47	13
Surprise	13	0	27	23	0	0	0	37

speech data set3, in which eight speakers' (four male and four female) speech utterances are used for training the models and two speakers' (one male and one female) data is used for the validation. The emotion recognition performance using set3 of IITKGP-SESC is given in Tables 3.4–3.12. Table 3.14 consolidates the emotion recognition performance using both databases (IITKGP-SESC and Emo-DB). AANNs are used for developing emotion classification models using (1) LP residual samples, (2) samples of LP residual phase, (3) epoch parameters, (4) epoch parameters at the syllable level, (5) samples of GVV signal and (6) parameters of the GVV signal. SVMs are used for developing emotion classification models using epoch parameters computed at the utterance level. In this work, nine emotion recognition systems (ERS-1(a) to ERS-7) are developed using different excitation source features mentioned in Sect. 3.4.

The emotion recognition system, developed using AANN models, is similar to the ER system developed using SVMs as shown in Fig. 3.11. An individual AANN model is developed for each of the emotions. For evaluating the performance of the ER system, the feature vectors derived from the test utterances are given as inputs to eight AANN models. Vector-wise output of each model is compared with its

Table 3.6 Emotion classification performance using samples of LP residual signal around glottal closure instants. Model: ERS-1(b) Average emotion recognition performance: 53.88

Emotions	Emotion recognition performance (%)							
	Anger	Disgust	Fear	Happiness	Neutral	Sadness	Sarcasm	Surprise
Anger	57	0	17	13	10	3	0	0
Disgust	0	54	0	3	13	17	3	10
Fear	0	23	60	0	0	7	3	7
Happiness	10	14	0	43	0	23	0	10
Neutral	0	10	0	0	63	27	0	0
Sadness	0	20	3	3	20	47	0	7
Sarcasm	0	23	0	0	17	3	50	7
Surprise	7	0	23	10	0	0	3	57

Table 3.7 Emotion classification performance using LP residual phase around glottal closure instants. Model: ERS-2(b) Average emotion recognition performance: 48.41

Emotions	Emotion recognition performance (%)							
	Anger	Disgust	Fear	Happiness	Neutral	Sadness	Sarcasm	Surprise
Anger	36	0	30	20	7	7	0	0
Disgust	0	70	0	0	7	13	0	10
Fear	13	20	27	10	0	20	0	10
Happiness	10	30	13	33	0	10	0	4
Neutral	0	3	0	0	77	10	0	10
Sadness	0	27	3	7	10	47	0	6
Sarcasm	0	33	7	0	10	0	50	0
Surprise	3	0	0	23	27	0	0	47

input y to compute the normalized squared error, e, which is given by: $e = \frac{\|y-o\|^2}{\|y\|^2}$, where o is the output vector given by the AANN model. The error e is transformed into a confidence score c, using; $c = exp(-e)$. The average confidence score per utterance is computed for each model using; $C_{avg} = \frac{1}{N} \left(\sum_{i=1}^{N} C_i \right)$ where N is the number of feature vectors per utterance. The model giving lowest error or the highest confidence measure is hypothesized as the recognized emotion.

In ERS-1(a), AANN models are expected to capture the higher order relations present among the samples of LP residual signal. In this task, 40 samples are considered to form a feature vector. The basic reason for choosing 40 samples is to capture the higher order relations present within a pitch period. Throughout this work, the speech signals with an 8 kHz sampling rate are used in conducting various studies. If the pitch period is less than 5 ms then, LP residual samples present in 75% of the pitch period are up-sampled to derive a feature vector of length 40. Feature vector size 40 indicates a block of 40 LP residual samples. The sequence of feature vectors is derived from LP residual signal by shifting the block by one sample each time. Each block of 40 samples (each feature vector) is normalized with respect to energy, before giving it as an input to the network. During normalization,

Table 3.8 Emotion classification performance using epoch parameters. Model: ERS-3 Average emotion recognition performance: 50

Emotions	Emotion recognition performance (%)							
	Anger	Disgust	Fear	Happiness	Neutral	Sadness	Sarcasm	Surprise
Anger	53	0	17	13	10	3	0	4
Disgust	0	50	0	4	13	23	0	10
Fear	3	23	57	10	0	7	0	0
Happiness	10	24	0	40	0	23	3	0
Neutral	0	10	0	0	60	20	3	7
Sadness	0	27	3	3	20	43	0	4
Sarcasm	0	23	10	0	0	0	50	17
Surprise	0	0	17	23	13	0	0	47

Table 3.9 Emotion classification performance using epoch parameters at syllable level. Model: ERS-4 Average emotion recognition performance: 48.38

Emotions	Emotion recognition performance (%)							
	Anger	Disgust	Fear	Happiness	Neutral	Sadness	Sarcasm	Surprise
Anger	43	0	27	13	10	0	3	7
Disgust	0	63	0	0	7	27	0	3
Fear	23	20	30	17	10	0	0	0
Happiness	17	20	7	37	0	0	0	23
Neutral	0	7	0	0	73	17	3	0
Sadness	0	33	0	0	10	40	7	10
Sarcasm	0	20	13	0	0	17	57	0
Surprise	3	3	30	17	0	0	0	47

Table 3.10 Emotion classification performance using samples of the glottal volume velocity signal. Model: ERS-5 Average emotion recognition performance: 51.25

Emotions	Emotion recognition performance (%)							
	Anger	Disgust	Fear	Happiness	Neutral	Sadness	Sarcasm	Surprise
Anger	53	0	17	13	7	10	0	4
Disgust	0	50	0	4	13	27	0	6
Fear	3	20	57	13	0	7	0	0
Happiness	10	24	0	40	0	23	3	0
Neutral	0	10	0	0	63	17	3	7
Sadness	0	23	7	3	20	43	0	4
Sarcasm	0	23	13	0	0	10	47	7
Surprise	0	0	10	20	13	0	0	57

each LP residual sample $r(n)$ is divided by the root-mean-squared value of the samples present in that frame. The normalization factor is given by $\sqrt{\frac{1}{N}\sum_{n=1}^{N} s(n)^2}$. Initially, the weights of the network are assigned to random values between -1 and $+1$. The network is trained with 100 epochs using a back-propagation algorithm. The choice of number of epochs is dictated by the training error. Block diagrams of

Table 3.11 Emotion classification performance using GVV parameters. Model: ERS-6 Average emotion recognition performance: 47

Emotions	Emotion recognition performance (%)							
	Anger	Disgust	Fear	Happiness	Neutral	Sadness	Sarcasm	Surprise
Anger	33	0	30	20	7	0	3	0
Disgust	0	67	0	0	10	20	0	3
Fear	17	20	33	13	0	17	0	0
Happiness	10	30	3	30	0	24	3	0
Neutral	0	10	0	0	70	20	0	0
Sadness	0	30	3	7	3	50	0	7
Sarcasm	0	20	17	0	10	13	43	0
Surprise	13	0	17	20	0	0	0	50

Table 3.12 Emotion classification performance using epoch parameters at utterance level. Model: ERS-7 Average emotion recognition performance: 52.13

Emotions	Emotion recognition performance (%)							
	Anger	Disgust	Fear	Happiness	Neutral	Sadness	Sarcasm	Surprise
Anger	53	0	13	17	17	0	0	0
Disgust	0	47	0	0	13	300	10	
Fear	17	0	60	23	0	0	0	0
Happiness	17	0	7	63	13	0	0	0
Neutral	0	10	0	0	57	23	0	10
Sadness	0	3	0	7	30	47	0	13
Sarcasm	0	33	0	0	10	17	40	0
Surprise	0	0	23	23	6	0	0	50

training and testing phases are shown in Fig. 3.12. The classification performance using the samples of LP residual is shown in Table 3.4, in the form of a confusion matrix. The network structure used for this experiment is 40L-60N-12N-60N-40L.

ERS-2(a) is developed using the phase of LP residual samples. The training and testing stages follow the same procedure as explained in the case of ERS-1(a). The results are given in Table 3.5.

In the LP residual signal, the region around the glottal closure instant, within each pitch cycle corresponds to a high SNR region, due to impulse-like excitation. This region is mostly the unpredictable part of the speech during LP analysis, whereas in glottal open regions, LP residual samples are random in nature and have very small amplitude. Hence, it may be hypothesized that the region around glottal closure contains crucial emotion-specific information. Therefore, only the blocks of LP residual signal around the GC region are explored separately for classifying the emotions. In this work, the instants of GC are determined using the zero frequency filtering method [130]. From the experiments, it is observed that about 16 blocks of LP residual samples around GC region are sufficient to capture the emotion-specific information. This approach reduces computational complexity and avoids the blocks of LP residual samples present in glottal open regions while deriving

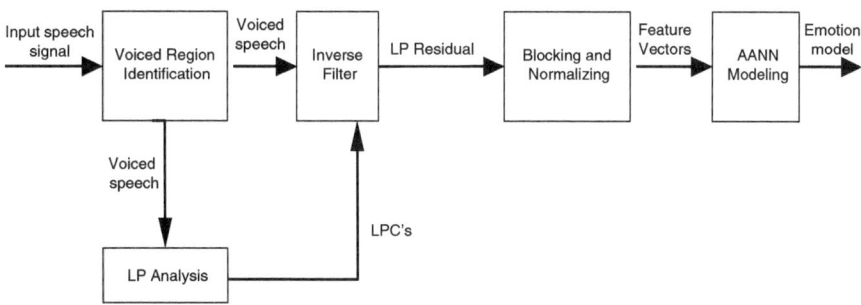

Training phase: Developing AANN models

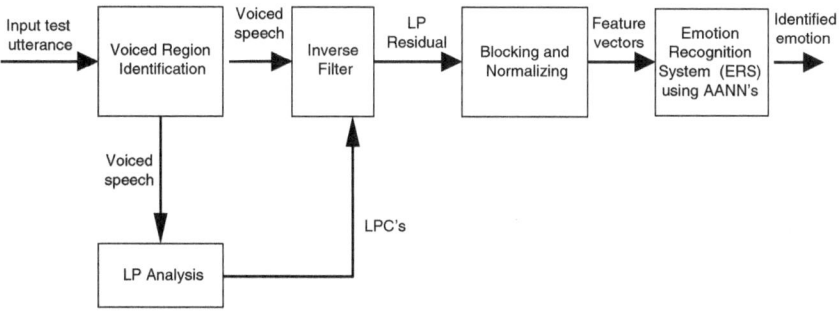

Testing phase: Validation of AANN models

Fig. 3.12   Training and testing phases of emotion recognition models using AANN

feature vectors for emotion classification. The performance of ER systems ERS 1-(b) and 2-(b), using the LP residual signal and its phase extracted around GC region, is given in Tables 3.6 and 3.7 respectively. The network structures used for these two experiments are the same as the network of ERS-1(a).

The recognition performance from Tables 3.6 and 3.7 indicates around 8% improvement over the performance shown in Tables 3.4 and 3.5. This may be due to the focusing on the crucial excitation source information and avoiding the noisy segments present in the glottal open regions.

ERS-3 is developed using the parameters derived from each epoch. The four epoch parameters constitute the four-dimensional feature vector for each epoch. Emotion recognition performance of ERS-3 is given in Table 3.8. The AANN structure used in this study is 4L-7N-3N-7N-4L.

ERS-4 is developed using the parameters of epoch sequence at the syllable level. Syllable boundaries are determined using vowel onset points (VOP) [151]. VOPs estimated using energies of LP residual, modulation spectrum and power spectrum are used as anchor points to decide syllable boundaries. The speech segment between two successive VOPs is basically considered as a syllable. The parameters of the epoch sequence present in the syllable are concatenated to form the feature vector. Twenty epochs are considered in each syllable. With this, the feature vector size becomes 80 for every syllable. The network structure used for building the model is 80L-160N-35N-160N-80L. The performance of the ERS-4 is given in Table 3.9.

ERS-5 is developed using the samples of the GVV signal. The higher order relations present in the glottal pulse signal are exploited to capture emotion-specific information. The emotion models of ERS-5 are developed in a similar manner as the models of ERS-1(a) are developed. Motivated by the improved results of Table 3.6 (than those of Table 3.4), feature vectors are extracted only around the GCI locations of the GVV signal. In each glottal cycle, 16 such blocks are considered to form the feature vectors. The AANN structure used to develop the models for ERS-5 is 40L-60N-12N-60N-40L. The main reason for using samples of the GVV signal is the difficulty in accurate estimation of its parameters. The performance of the model is shown in Table 3.10.

ERS-6 is developed using GVV parameters. The following parameters are computed for each pitch cycle of the GVV signal (see Fig. 3.8). (1) The time difference between the beginning of opening phase to the complete open phase gives the *duration of opening phase* $(t_{og} = t_1 - t_0)$. (2) Time difference between the complete open phase to the beginning of the closed phase gives *the duration of closing phase* $(t_{cg} = t_2 - t_1)$. (3) The time difference between the beginning of closed phase to the beginning of opening phase gives *the duration of complete closure* $(t_c = t_3 - t_2)$. (4) The *slope of opening phase* is determined as the ratio of magnitude difference of the first and the last samples of opening phase to the duration of the opening phase $(S_c = \frac{|a_1 - a_0|}{t_{cg}})$. (5) *The slope of closing phase* is determined as the ratio of magnitude difference of the first and the last samples of the closing phase to its duration $(S_o = \frac{|a_1 - a_0|}{t_{og}})$. (6) The ratio of slope of opening phase to that of closing phase is also computed as one of the parameters $(S_r = \frac{S_o}{S_c})$. The network structure used in this study is 6L-15N-4N-15N-6L. The performance of the ERS-6 is shown in Table 3.11.

ERS-7 is developed using parameters of the epoch sequence at the utterance level. SVMs are used for capturing discriminative information present in utterance level feature vectors. The differentiating capability of SVMs mainly depends upon discriminating feature vectors, known as support vectors, than the actual number of feature vectors. Here feature vectors are obtained by concatenating the epoch parameters of the sequence of epochs present in the utterance. For deriving the fixed dimensional feature vectors, 50 epochs per utterance are chosen from a sequence of epochs. The epoch parameters of these 50 epochs are concatenated to form a feature vector of size 200 (4 *epoch parameters* × 50 *epochs*). The emotion recognition performance using utterance level epoch parameters is given in Table 3.12.

From these studies (ERS-1(a) to ERS-7), it is observed that emotion-specific information is present in all proposed excitation source features. Some features have more discriminating power for certain emotions. For instance fear is identified

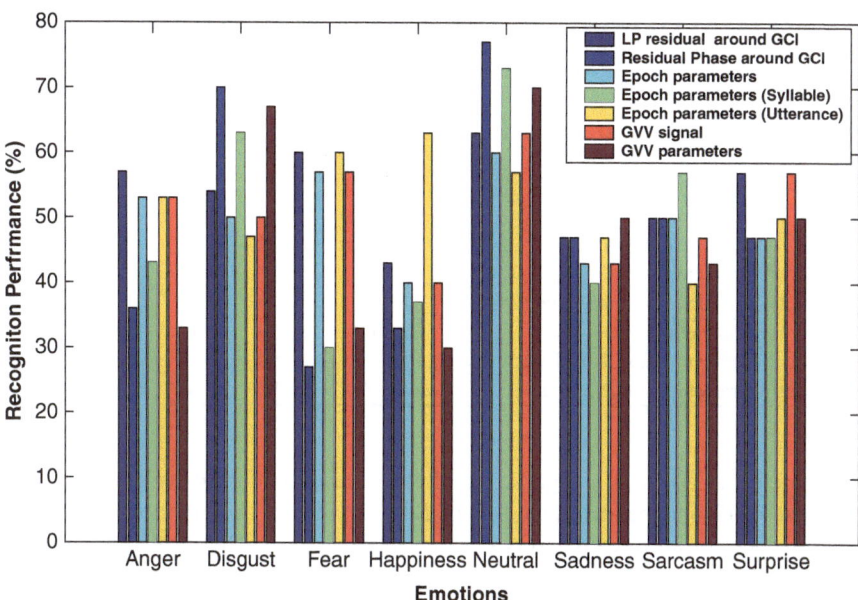

Fig. 3.13 Comparison of emotion recognition performance of different proposed excitation source features with respect to individual emotions

better using LP residual samples taken around glottal closure instants (ERS-1(b)) and the dynamics of epoch parameters derived from the utterances (ERS-7), with the recognition rate of 60%. Happiness is recognized better using the dynamics of utterance level epoch parameters (ERS-7). The LP residual phase extracted around GCI regions have performed well in the case of disgust and neutral emotions. The LP residual samples chosen around GCI regions (ERS-2(b)) has discriminated anger in a better way compared to other features. LP residual samples around GCI regions have recognized most of the emotions with more than 50% accuracy. Overall emotion recognition performance using LP residual phase is observed to be slightly lower compared to other excitation source (ES) features. This indicates that, LP residual phase may not have sufficient emotion specific information. Average recognition performance using epoch parameters at various levels is observed to be around 50%. Among these parameters, epoch parameters from the sequence of epochs present in the utterance have shown slightly better performance due to the presence of emotion specific dynamic information at utterance level. Among the excitation source features derived from GVV signal, the signal samples of GVV waveform have shown slight improvement in the performance over the other GVV parameters. Lower performance of GVV parameters may be due to their inaccurate estimation. From the results presented in the Tables 3.4–3.12 one of the unique observations is that, miss-classification of other emotions as sarcasm is rare. The details of emotion recognition with respect to each emotion using different excitation source features are presented in the bar chart shown in Fig. 3.13.

Table 3.13 Average emotion classification performance on Set1, Set2, and Set3 datasets of IITKGP-SESC using different excitation source features

Features	Models	Recognition performance (%)		
		Set1	Set2	Set3
LP residual signal	ERS-1(a)	51.62	47.92	46.38
LP residual samples around GCI	ERS-1(b)	59.92	57.61	53.88
LP residual phase	ERS-2(a)	47.63	45.13	41.38
LP residual phase around GCI	ERS-2(b)	55.72	51.02	48.41
Epoch parameters	ERS-3	56.81	53.46	50.00
Samples of glottal volume velocity signal	ERS-5	57.13	54.25	51.25
GVV parameters	ERS-6	54.00	50.13	47.00
Epoch parameters at syllable level	ERS-4	54.98	51.67	48.38
Epoch parameters at utterance level	ERS-7	57.82	55.45	52.13

The similar emotion recognition studies are also conducted on session (Set1) and text (Set2) independent data sets of IITKGP-SESC. The emotion recognition performance for all three datasets is given in Table 3.13. The results obtained on Set2 are slightly better compared to the results of Set3. This may be due to influence of speaker specific information during emotion recognition. The improvement is about 2–3% compared to the speaker and text independent results. The emotion recognition performance using Set1 dataset is observed to be better compared to the results of Set2 and Set3 datasets. This is mainly due to the bias of speech and text dependent emotion specific information. From the results, it indicates that, emotion recognition performance using Set1 has 3–4% more than the results of Set2 and 4–7% more than the results of Set3. The observations of the above discussion may be visualized in the Fig. 3.14.

The same set of studies have been conducted on Emo-DB and the results are shown in Table 3.14. From the results it is observed that the average emotion recognition performance on IITKGP-SESC is slightly better compared to the performance derived on Berlin emotion speech corpus, due to the larger size of IITKGP-SESC. These studies are conducted basically to evaluate the performance of the proposed features on other language. This helps to analyze language independent nature of the features during emotion classification. From the results (see Table 3.14), it is evident that the performance of emotion recognition models is not much affected by the language factor in the case of proposed excitation source features. The comparison of emotion recognition performance using IITKGP-SESC and Emo-DB, using the proposed ES features is shown in the Fig. 3.15.

The aim of this work is to show that, along with the spectral and prosodic components of speech, excitation source also carries some emotion-specific information. From the studies presented in this work, it is observed that, the proposed excitation source features will not produce a classification performance comparable to the state of the art literature on their own. However it is also observed that the measure provided by the proposed excitation source features may be supplementary to the conventional VT system and prosodic features.

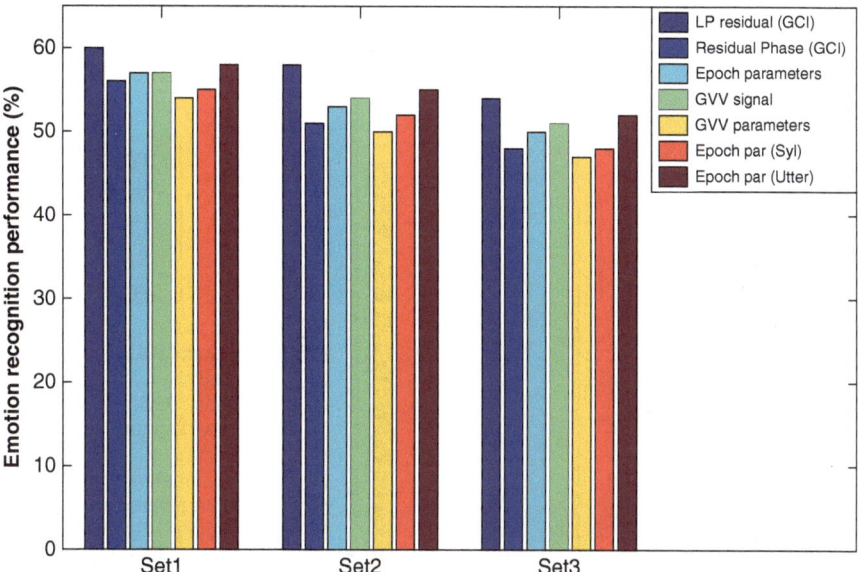

Fig. 3.14 Comparison of emotion recognition performance of different proposed excitation source features with respect to Set1, Set2, and Set3 of IITKGP-SESC

Table 3.14 Average emotion classification performance on IITKGP-SESC and Berlin emotional speech corpus using different excitation source features

		Rec. per (%)	
Features	Models	IITKGP SESC	Berlin database
LP residual signal	ERS-1(a)	46.38	44.33
LP residual samples around GCI	ERS-1(b)	53.88	52.43
LP residual phase	ERS-2(a)	41.38	39.16
LP residual phase around GCI	ERS-2(b)	48.41	41.33
Epoch parameters	ERS-3	50.00	48.00
Samples of glottal volume velocity signal	ERS-5	51.25	49.66
GVV parameters	ERS-6	47.00	47.26
Epoch parameters at syllable level	ERS-4	48.38	45.33
Epoch parameters at utterance level	ERS-7	52.13	51.45

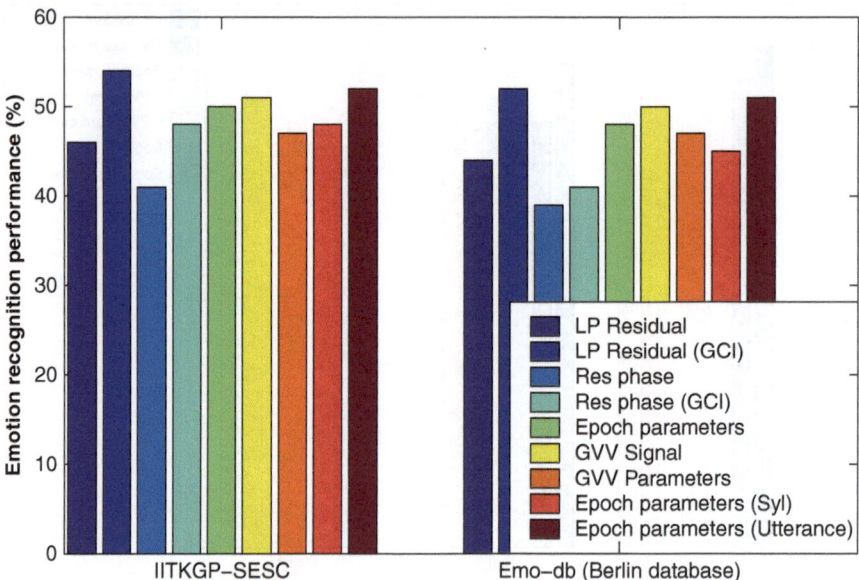

Fig. 3.15 Comparison of emotion recognition performance of different proposed excitation source features with respect to IITKGP-SESC and Emo-DB

3.7 Summary

In this chapter, features extracted from the excitation source signal are explored for characterizing and categorizing the emotions present in speech.

Subjective listening tests, conducted on emotional speech utterances with respect to excitation source, vocal tract system and their combination, have demonstrated that the emotion-specific information is present in each component of the speech. The following features, extracted from the excitation source signal, are proposed for characterizing the emotions: (a) LP residual samples, (b) phase of the LP residual samples, (c) epoch parameters, (d) dynamics of the epoch parameters at syllable and utterance level, (e) samples of the GVV signal and (f) parameters of the GVV signal. The Berlin emotional speech corpus (Emo-DB) and Telugu emotional database (IITKGP-SESC) are used to carry out these studies. Out of several excitation source features proposed, LP residual samples and epoch parameters extracted around glottal closure instants are found to have better emotion discrimination characteristics. AANN models are used for capturing emotion-specific information from the LP residual, phase of the LP residual, epoch parameters, and GVV parameters. SVM models are used for discriminating the emotions using epoch parameters at the syllable and utterance level. The performance of emotion recognition systems developed using different excitation source features vary between 39% and 54%. From the results, it is observed that each of the proposed excitation source features has considerable contribution toward discriminating the emotions.

Chapter 4
Emotion Recognition Using Vocal Tract Information

Abstract This chapter discuss about the emotion specific information offered by vocal tract features. Well known spectral features such as linear prediction cepstral coefficients (LPCCs) and mel frequency cepstral coefficients (MFCCs) are used as the correlates of vocal tract information for discriminating the emotions. In addition to LPCCs and MFCCs, formant related features are also explored in this work for recognizing emotions from speech. Extraction of the above mentioned spectral features is discussed in brief. In this study, auto-associative neural network (AANN) models and Gaussian mixture models (GMM) are used for classifying the emotions. Functionality of AANN and GMM are briefly described. Emotion recognition performance obtained using different vocal tract features are compared over Indian and Berlin emotional speech databases. Performance of neural networks and Gaussian mixture models in classifying the emotional utterances based on vocal tract features is also evaluated.

4.1 Introduction

In Chap. 3, we have discussed the recognition of emotions using excitation source features. This study indicates that the excitation source signal contains some useful emotion-specific information, however, these features alone may not be sufficient to provide appreciably high emotion recognition performance. The performance of the emotion recognition systems may be improved by combining excitation source information with the spectral (vocal tract) features. In this chapter, we intend to discuss the use of vocal tract system features for speech emotion recognition.

Generally, spectral features are found to be robust for various speech tasks [152]. This may be due to the accurate representation of vocal tract system characteristics using spectral features. Figure 4.1 shows the unique spectral characteristics for eight emotions of IITKGP-SESC. The spectra shown in Fig. 4.1 represent the steady region of a vowel /A/ from Telugu utterance *anni d$\underline{A}$namulalo vidyA dAnamu minnA*, expressed in eight different emotions. It is observed from the figure that

S.R. Krothapalli and S.G. Koolagudi, *Emotion Recognition using Speech Features*,
SpringerBriefs in Electrical and Computer Engineering,
DOI 10.1007/978-1-4614-5143-3_4, © Springer Science+Business Media New York 2013

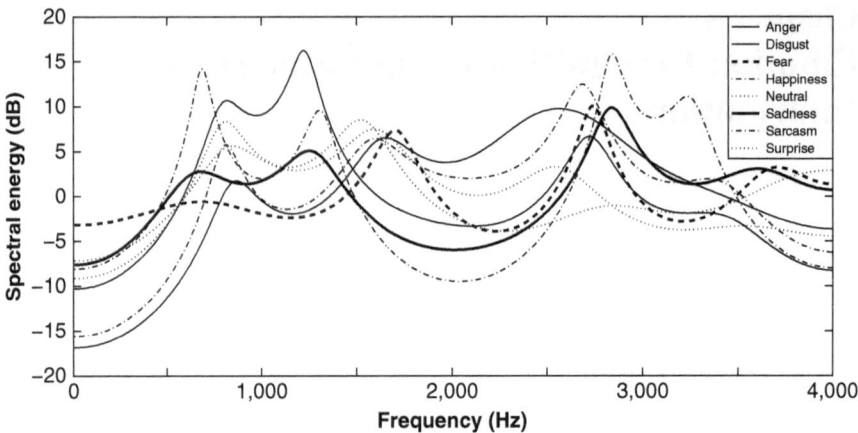

Fig. 4.1 Spectra of a steady region of vowel /A/, taken from the utterance *anni dAnamulalo vidyA dAnamu minnA*, in eight emotions

the sharpness of the formant peaks, positions of the formants, formant bandwidths, and spectral tilt have distinctive properties for different emotions. In the literature, spectral features are used for modeling the pitch pattern of the speaker [91]. Variation of the pitch is an important correlate of emotions in speech. It is also known that variation in the pitch leads to changes in other prosodic parameters like duration and energy. Intuitively it may be harder to correlate spectral features with the temporal dynamics in prosody, related to emotional states. But they provide a detailed description of varying speech signal, including variations in prosody [92]. Our intuition here is that spectral features capture not only the information about what is being said (text) but also how it is being said (emotion). Therefore, spectral features are used in this study to recognize speech emotions.

The majority of the works on speech emotion recognition are carried out using spectral features derived from the whole speech signal through a conventional block processing approach. In this chapter, the same conventional spectral features are used for identifying the speech emotions.

The remaining part of the chapter is organized as follows, Extraction of different types of spectral features mentioned above is discussed in Sect. 4.2. Section 4.3 discusses the methodology of Gaussian mixture models used to develop emotion recognition models. Emotion recognition results are discussed in Sect. 4.4. Chapter concludes with a summary.

4.2   Feature Extraction

In this work, LPCCs, MFCCs and formant related features are used for representing the spectral information to recognize emotions. Extraction of different spectral features, mentioned above is discussed in the following subsections.

4.2.1   *Linear Prediction Cepstral Coefficients (LPCCs)*

The cepstral coefficients derived from either linear prediction (LP) analysis or a filter bank approach are almost treated as standard front end features. Speech systems developed based on these features have achieved a very high level of accuracy, for speech recorded in a clean environment. Basically, spectral features represent phonetic information, as they are derived directly from spectra. The features extracted from spectra, using the energy values of linearly arranged filter banks, equally emphasize the contribution of all frequency components of a speech signal. In this context, LPCCs are used to capture emotion-specific information manifested through vocal tract features. In this work, the tenth order LP analysis has been performed, on the speech signal, to obtain 13 LPCCs per speech frame of 20 ms using a frame shift of 10 ms. The human way of emotion recognition depends equally on two factors, namely: its expression by the speaker as well as its perception by a listener. The purpose of using LPCCs is to consider vocal tract characteristics of the speaker, while performing automatic emotion recognition.

Cepstrum may be obtained using linear prediction analysis of a speech signal. The basic idea behind linear predictive analysis is that the nth speech sample can be estimated by a linear combination of its previous p samples as shown in the following equation.

$$s(n) \approx a_1 s(n-1) + a_2 s(n-2) + a_3 s(n-3) + \cdots + a_p s(n-p)$$

where a_1, a_2, a_3 $\cdots$ are assumed to be constants over a speech analysis frame. These are known as predictor coefficients or linear predictive coefficients. These coefficients are used to predict the speech samples. The difference of actual and predicted speech samples is known as an error. It is given by

$$e(n) = s(n) - \hat{s}(n) = s(n) - \sum_{k=1}^{p} a_k s(n-k)$$

where $e(n)$ is the error in prediction, $s(n)$ is the original speech signal, $\hat{s}(n)$ is a predicted speech signal, $a_k s$ are the predictor coefficients.

To compute a unique set of predictor coefficients, the sum of squared differences between the actual and predicted speech samples has been minimized (error minimization) as shown in the equation below

$$E_n = \sum_{m} \left[s_n(m) - \sum_{k=1}^{p} a_k s_n(m-k) \right]^2$$

where m is the number of samples in an analysis frame. To solve the above equation for LP coefficients, E_n has to be differentiated with respect to each a_k and the result is equated to 0 as shown below

$$\frac{\partial E_n}{\partial a_k} = 0, \qquad \text{for} k = 1, 2, 3, \cdots, p$$

After finding the a_ks, one may find cepstral coefficients using the following recursion.

$$C_0 = \log_e p$$

$$C_m = a_m + \sum_{k=1}^{m-1} \frac{k}{m} C_k a_{m-k}, \qquad \text{for } 1 < m < p \text{ and}$$

$$C_m = \sum_{k=m-p}^{m-1} \frac{k}{m} C_k a_{m-k}, \qquad \text{for } m > p$$

4.2.2   Mel Frequency Cepstral Coefficients (MFCCs)

A human auditory system is assumed to process a speech signal in a nonlinear fashion. It is well known that lower frequency components of a speech signal contain more phoneme specific information. Therefore a nonlinear mel scale filter bank has been used to emphasize lower frequency components over higher ones. In speech processing, the mel frequency cepstrum is a representation of the short term power spectrum of a speech frame using a linear cosine transform of the log power spectrum on a nonlinear mel frequency scale. Conversion from normal frequency f to mel frequency m is given by the equation

$$m = 2{,}595 \log_{10} \left(\frac{f}{700} + 1 \right)$$

The steps used for obtaining mel frequency cepstral coefficients (MFCCs) from a speech signal are as follows:

1. Pre-emphasize the speech signal.
2. Divide the speech signal into a sequence of frames with a frame size of 20 ms and a shift of 5 ms. Apply the hamming window over each of the frames.
3. Compute the magnitude spectrum for each windowed frame by applying DFT.
4. Mel spectrum is computed by passing the DFT signal through a mel filter bank.
5. DCT is applied to the log mel frequency coefficients (log mel spectrum) to derive the desired MFCCs.

Twenty filter banks are used to compute 8, 13 and 21 MFCC features from a speech frame of 20 ms, with 5 ms overlap each time. The purpose of using MFCCs is to take the listener's non-linear auditory perceptual system into account, while performing automatic emotion recognition. More details about MFCC feature extraction are given in Appendix B.

4.2.3   Formant Features

Cepstral coefficients are used as standard front end features for developing various speech systems, however, they perform poorly with noisy or real life speech. Therefore the supplementary features along with basic cepstral coefficients are essential to handle real life speech. The higher amplitude regions of a spectrum, such as formants, are relatively less affected by the noise. K.K. Paliwal et al. have extracted spectral sub-band centroids from high amplitude regions of the spectrum and used for noisy speech recognition [153]. With this viewpoint, formant parameters are used in this study as the supplementary features to cepstral features. Also note that the conventional cepstral features utilize only amplitude (energy) information from the speech power spectrum, whereas the proposed formant features utilize frequency information as well.

In general, formant tracks represent the sequences of vocal tract shapes, hence formant analysis using their strength, location and bandwidth may help to extract vocal tract related emotion specific information from a speech signal. Figure 4.2 shows different spectra for eight emotions of IITKGP-SESC. The spectra are derived from the syllable *tha* from Telugu sentence *thallidhandrulanu gauravincha valenu*. In this case, the language, text, speaker and contextual information is maintained the same. This is speculative from the figure that the variation in the spectra is due to the emotions. Formant frequencies are very crucial in view of speech perception. Hence, a slight change in these parameters causes a perceptual difference, which may lead to manifestation of different emotions. It is evident from Fig. 4.2 that the position and strength of formants are clearly distinct for different emotions. Spectral peaks indicate the intensity of specific frequency components (or frequency band). Their distinctive nature for different emotions is the indication of presence of emotion specific information. The rate of decrease in spectrum amplitude, as a function of frequency is known as spectral roll-off or spectral tilt. This happens mainly because of decreasing strength of harmonics, as the frequency increases. A speaker can induce more strength into higher harmonics by consciously controlling the glottal vibration. Abrupt closing of the glottis increases the energy in the higher frequency components. This leads to the variation in spectral roll-off for different emotions. Figure 4.2 shows distinct spectral roll-offs for each of the emotions.

Though it is assumed that the bandwidth of a formant does not influence phonetic information [15], it represents some speaker specific information. Figure 4.2 depicts the variation in the formant bandwidths in the case of different emotions. Even a slight variation in the bandwidth may be due to speaker induced emotion specific information as speaker, text, language and context related information do remain the same. Formant bandwidth is the frequency band measured at around 3 dB downward from the respective formant peak.

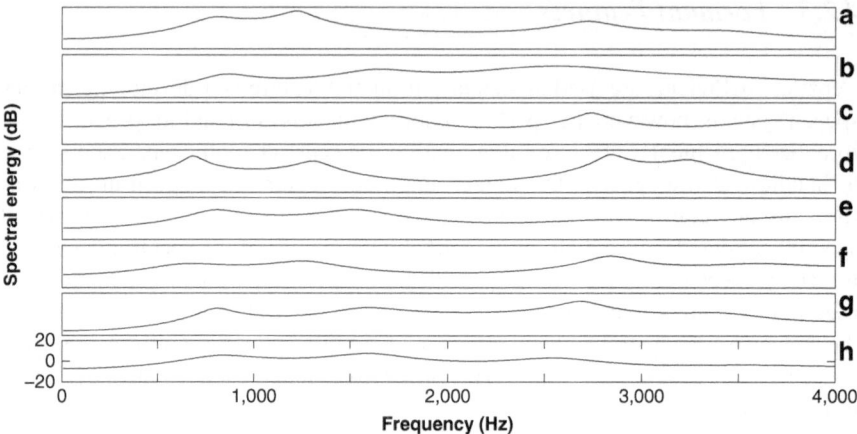

Fig. 4.2 Spectra for the steady region of the syllable /tha/ from the utterance *thallidhandrulanu gauravincha valenu*. (**a**) Anger, (**b**) Disgust, (**c**) Fear, (**d**) Happiness, (**e**) Neutral, (**f**) Sadness, (**g**) Sarcasm, and (**h**) Surprise

4.3 Classifiers

GMMs and AANNs are known to capture the general distribution of data points in the feature space and one can be used as an alternative to the other [154]. Two classifiers are used in this study, to mutually compare their emotion classification results. The philosophy of auto-associative neural networks is discussed in the Sect. 3.5 of Chap. 3.

4.3.1 Gaussian Mixture Models (GMM)

Gaussian Mixture Models (GMMs) are among the most statistically mature methods for clustering. A Gaussian mixture model is used as a classification tool in this task. They model the probability density function of observed variables using a multivariate Gaussian mixture density. Given a series of inputs, a GMM refines the weights of each distribution through the expectation-maximization algorithm. Mixture models are a type of density model, which comprise a number of component functions, usually Gaussian in nature. These component functions are combined to provide a multi-modal density. Mixture models are a semi-parametric alternative to non-parametric models and provide greater flexibility and precision in modeling the underlying statistics of sample data. They are able to smooth over gaps resulting from sparse sample data and provide tighter constraints in assigning object membership to cluster regions. Once a model is generated, conditional probabilities can be computed for test patterns (unknown data points). An expectation maximization

(EM) algorithm is used for finding maximum likelihood estimates of parameters in probabilistic models, where the model depends on unobserved latent variables.

Expectation Maximization is an iterative method that alternates between performing an expectation (E) step, which computes an expectation of the log likelihood with respect to the current estimate of the distribution for the latent variables, and a maximization (M) step, which computes the parameters that maximize the expected log likelihood found on the E step. These parameters are then used to determine the distribution of the latent variables in the next E step.

The number of Gausses in the mixture model is also known as the number of components. They indicate the number of clusters in which data points are to be distributed in order to cover local variations. In this work, one GMM is developed to capture the information about one emotion. Depending on the number of training data points, the number of components may be varied in each GMM. The presence of few components in a GMM, and training it with large number of data points may lead to more generalized clusters, failing to capture specific details related to each class. On the other hand over-fitting of the data points may happen, if too many components represent few data points. Obviously the complexity of the model increases, if they contain higher numbers of components. Therefore a tradeoff has to be reached between the complexity and the accuracy of the classification results required. In this work, GMMs are designed with 64 components and iterated 30 times to attain convergence. A diagonal covariance matrix is used to derive the model parameters. The philosophy and principles of Gaussian mixture models are discussed in detail in Appendix C.

4.4   Results and Discussion

In this work, five emotion recognition systems are developed to study speech emotion recognition using different spectral features. In the beginning, emotion recognition systems are developed individually, using MFCCs, LPCCs, and formant features. Formant features alone have not given appreciably good emotion recognition performance, therefore, in the later stages, they are used in combination with the other features. In the following paragraphs, the emotion recognition performance of all individual emotion recognition systems, developed using Set3 of IITKGP-SESC, are discussed. Out of ten speakers' speech data, the utterances of eight speakers (four male and four female) are used for training the ER models and the utterances of two (a male and a female) speakers are used for validating the trained models. Thirteen spectral features are extracted from a frame of 20 ms, with a shift of 5 ms. GMMs with 64 components are used to develop ERSs. The results of emotion recognition performance using session and text independent (Set1 and Set2) speech data are also given at the end of the chapter.

ERS1 is developed using 13 LPCC features obtained from the entire speech signal using the normal block processing approach. The emotion recognition performance of ERS1 is shown in Table 4.1. Diagonal elements of the table

Table 4.1 Emotion classification performance, using LPCC features (ERS1). Average emotion recognition: 68%

	Emotion recognition performance (%)							
Emotions	Anger	Disgust	Fear	Happiness	Neutral	Sadness	Sarcasm	Surprise
Anger	53	3	17	13	7	0	0	7
Disgust	0	63	0	0	20	17	0	0
Fear	10	0	67	13	3	0	0	7
Happiness	7	0	13	77	0	0	0	3
Neutral	0	7	0	0	80	13	0	0
Sadness	0	10	0	0	17	70	0	3
Sarcasm	0	7	0	0	10	0	73	10
Surprise	10	0	17	13	0	0	0	60

Table 4.2 Emotion classification performance using MFCC features (ERS2). Average emotion recognition: 63.38%

	Emotion recognition performance (%)							
Emotions	Anger	Disgust	Fear	Happiness	Neutral	Sadness	Sarcasm	Surprise
Anger	57	0	17	23	0	0	0	7
Disgust	0	53	0	3	20	17	0	7
Fear	10	0	60	27	0	0	0	3
Happiness	7	0	23	70	0	0	0	0
Neutral	0	10	0	0	77	13	0	0
Sadness	0	7	0	0	27	63	0	3
Sarcasm	0	13	0	0	0	17	70	0
Surprise	7	0	13	23	0	0	0	57

show correct classification and the other elements indicate the percentage of misclassified utterances. ERS2 is developed using 13 MFCC features extracted frame wise from entire speech signal. Table 4.2 shows the corresponding emotion recognition performance. ERS3 is developed using formant related features. These 13 formant related features (4 frequencies, 4 energy values, 4 bandwidth values and a slope), extracted per frame of 20 ms, are used to represent formant information. Their concatenation forms the 13 dimensional feature vector. Table 4.3 shows the emotion recognition performance using only formant features. The average emotion recognition performance using formant features is about 47%.

ERS4 and ERS5 are developed using the combination of 13 formant features along with 13 LPCCs and 13 MFCCs respectively. The dimension of the resulting feature vectors is 26. Tables 4.4 and 4.5 show the emotion recognition performance of ERS4 and ERS5 respectively.

The results of the above studies show that emotion recognition performance using LPCCs is better than the results of MFCCs. The reason for this may be that LPCCs mainly represent the speech production characteristics, by analyzing all frequency components in a uniform manner. The emotion specific information may be present across all the frequencies in a uniform way. The proposed formant features alone are

Table 4.3 Emotion classification performance using formant features (ERS3). Average emotion recognition: 47%

Emotions	Emotion recognition performance (%)							
	Anger	Disgust	Fear	Happiness	Neutral	Sadness	Sarcasm	Surprise
Anger	33	0	7	3	30	0	27	0
Disgust	0	33	0	0	30	10	27	0
Fear	3	0	37	0	30	0	30	0
Happiness	7	0	3	47	23	0	20	0
Neutral	0	0	0	0	66	17	10	7
Sadness	0	10	0	0	17	57	13	3
Sarcasm	0	13	0	0	13	0	64	10
Surprise	0	0	0	0	37	0	23	40

Table 4.4 Emotion classification performance using the combination of LPCC + formant features (ERS4). Average emotion recognition: 69%

Emotions	Emotion recognition performance (%)							
	Anger	Disgust	Fear	Happiness	Neutral	Sadness	Sarcasm	Surprise
Anger	60	0	23	17	0	0	0	0
Disgust	0	60	0	7	17	13	0	3
Fear	7	0	63	17	10	0	0	3
Happiness	10	0	17	73	0	0	0	0
Neutral	0	0	0	0	83	17	0	0
Sadness	0	10	0	0	10	73	0	7
Sarcasm	0	0	0	0	13	3	77	7
Surprise	3	7	10	17	0	0	0	63

Table 4.5 Emotion classification performance using the combination of MFCC+formant features (ERS5). Average emotion recognition: 68%

Emotions	Emotion recognition performance (%)							
	Anger	Disgust	Fear	Happiness	Neutral	Sadness	Sarcasm	Surprise
Anger	63	0	3	17	10	0	0	7
Disgust	0	60	0	0	23	7	0	10
Fear	13	0	63	17	7	0	0	0
Happiness	10	0	3	70	0	0	0	17
Neutral	0	0	0	0	77	13	0	10
Sadness	0	10	0	0	17	73	0	0
Sarcasm	0	17	0	0	13	0	60	10
Surprise	0	10	0	0	17	20	0	53

not suitable to develop emotion recognition systems as their individual performance is poor. However, the combination of formant features with other spectral features has been proved to improve the recognition performance.

Different numbers of LPCCs/MFCCs are also explored for analyzing the emotion recognition performance. Table 4.6 indicates the average emotion recognition performance of 8 emotions using 8, 13 and 21 spectral features. It may be observed

Table 4.6 Average emotion classification performance on IITKGP-SESC using varying number of cepstral coefficients

	No. of ceps. coeff.		
	8	13	21
Features	Recognition performance (%)		
LPCCs	59	68	69
MFCCs	55	63	65
Formant features	47		
LPCCs + formants	58	69	**70**
MFCCs + formants	57	65	67

Table 4.7 Average emotion classification performance of GMM and AANN models using spectral features on IITKGP-SESC

	GMMs	AANNs	AANN structure
Features	Rec. per. (%)		
LPCCs	69	**63**	21-45-10-45-21
MFCCs	65	59	21-45-10-45-21
Formant features	47	41	13-28-7-28-13
LPCCs + formants	**70**	61	34-60-15-60-34
MFCCs + formants	67	60	34-60-15-60-34

from Table 4.6 that most of the times, the systems developed using higher numbers of spectral features have performed slightly better than their counterparts developed using smaller numbers of spectral features. It may be due to the reason that higher order spectral features contain more specific information about paralinguistic aspects of the speech, such as speaker, rhythm, melody, timbre, emotion and so on [87].

Auto-associative neural networks are known for capturing the non-linear relations among the feature vectors [155]. AANNs are also capable of capturing the distribution properties as GMMs do [154]. So the best performance by GMMs, using 21 spectral features, is compared with the results of relevant AANNs. AANNs are used as the emotion classifiers to cross validate empirically the results obtained by GMMs. Table 4.7 shows the comparison of emotion recognition results of both GMMs and AANNs. AANN structures used for developing emotion models are also given in the last column of the table. From the results, it is observed that emotion recognition performance using GMMs is better than that of AANN models. This indicates that emotion specific information from the spectral features is better captured by GMMs than by the AANNs. The proposed spectral features are also tested on internationally known Berlin emotion speech corpus (Emo-DB). The results obtained are on par with the results of our Indian database (IITKGP-SERSC). Table 4.8 shows the comparison of emotion recognition results obtained using IITKGP-SESC and Emo-DB.

The results given in Table 4.8 are obtained using 21 spectral and 13 formant features. The emotion recognition systems are developed using GMMs. From the

Table 4.8 Average emotion classification performance using the proposed spectral features on IITKGP-SESC and Emo-DB

Features	IITKGP-SESC	Emo-DB
	Rec. per. (%)	
LPCCs	69	64
MFCCs	65	63
Formant features	47	41
LPCCs + formants	**70**	**68**
MFCCs + formants	67	63

Table 4.9 Average emotion classification performance using Set1, Set2, and Set3 datasets of IITKGP-SESC

Features	Recognition performance (%)		
	Set1	Set2	Set3
LPCCs	74	71	69
MFCCs	67	65	65
Formant features	53	49	47
LPCCs + formants	**75**	**72**	**70**
MFCCs + formants	71	68	67

table, it is evident that the trends of emotion recognition performance using different spectral features on IITKGP-SESC and Emo-DB are almost the same.

So far we have carried out emotion recognition using the Set3 data set of IITKGP-SESC, which represents speaker and text independent emotion recognition. Along with Set3, emotion recognition results of Set1 and Set2 are given in Table 4.9. Recognition performance of Set2 data set is 2% higher than the results of Set3 data set. This is mainly due to the influence of speaker specific information during emotion classification. In the case of Set1, the recognition performance is about 4%–5% more than the results of Set3 and about 2%–3% more than the results of Set2. This improvement in emotion recognition is due to text and speaker specific information.

4.5 Summary

In this chapter, basic spectral features directly derived from entire speech utterances through conventional block processing have been used for emotion recognition. IITKGP-SESC and Emo-DB are used to carry out the emotion classification using the proposed spectral features. LPCCs, MFCCs and formant features are used as features to represent vocal tract information. The studies conducted in this chapter indicate that spectral features contain more discriminating properties with respect to different emotions than excitation source features. In the literature, MFCCs are claimed to be robust features for majority of the speech tasks such as: speech recognition and synthesis. Surprisingly, in this study, LPCCs are found to be outperforming MFCCs while classifying the emotions. Formant features combined with basic spectral features have always improved the emotion recognition performance of the systems by a consistent margin of around 2%–3%. Two classification models, namely GMMs and AANNs, are used for developing emotion recognition systems. The majority of the results reported in this chapter are obtained using GMMs as they performed slightly better than AANNs.

Chapter 5
Emotion Recognition Using Prosodic Information

Abstract In this chapter prosodic features are investigated for in view of discriminating the emotions from speech. The motivation for exploring the prosodic features to recognize the emotions is illustrated using the gross statistics and time varying patterns of prosodic parameters. Global prosodic features representing the gross statistics of prosody and local prosodic features representing the finer variations in prosody are introduced in this chapter for discriminating the emotions. Extraction procedures of global and local prosodic features are briefly discussed. In this study, support vector machines are used for capturing the emotion-specific information from the proposed global and local prosodic features. Performance of the developed emotion recognition systems are analyzed with respect to individual components of prosody and their combinations.

5.1 Introduction

In the literature, prosodic features are treated as major correlates of vocal emotions. The effect of emotions on basic prosodic parameters such as pitch, energy, and duration is analyzed in several studies (refer to Table 2.4). Human beings also mostly use the prosodic cues for identifying the emotions present in day-to-day conversations. For instance pitch and energy values are high for active emotions like anger, whereas the same parameters are comparatively lower for the passive emotion like sadness. Duration used for expressing anger is shorter than the duration used for sadness. Existing works on the use of prosody for emotion recognition have mainly focused on static prosodic values (mean, maximum, minimum and so on) for discriminating the emotions. However, the time varying dynamic nature of prosodic contours seems to be more emotion specific. According to the literature presented in Sect. 2.5, there is little evidence on using the dynamic nature of prosodic contours for speech emotion recognition. In this chapter, static and dynamic prosodic features extracted at utterance level are analyzed to study their contribution towards speech emotion recognition. The study on these issues is carried out by performing the following tasks: (1) Analysis of speech emotions using static prosodic features, (2) Investigating dynamic prosodic features at the utterance level, for discriminating the emotions, and (3) Combination of measures due to static (global) and dynamic (local) prosodic features at utterance level to recognize the emotions. Support vector machine (SVM) models are used for developing the emotion recognition models for discriminating the emotions. Statistical parameters like mean, minimum, maximum, standard deviation, derived from the sequence of prosodic values, are known as global or static features. The parameters representing temporal dynamics of the prosodic contours are known as local or dynamic features. In this work, the terms, static and global, dynamic and local are used as synonyms.

The rest of the chapter is organized as follows: Section 5.2 discusses the importance of prosodic features in classifying the speech emotions. Section 5.3 briefly mentions the motivation behind this study. Section 5.4 explains the details of extraction of static and dynamic prosodic features. Evaluation of the developed emotion recognition models and their performance are discussed in Sect. 5.5. A brief summary of the present work is given in Sect. 5.6.

5.2   Prosodic Features: Importance in Emotion Recognition

Normally human beings use dynamics of long term speech features like energy profile, intonation pattern, duration variations and formant tracks, to process and perceive the emotional content from the speech utterances. This might be the main reason for the extensive use of prosodic features by most of the research community. However, many times humans tend to get confused while distinguishing the emotions that share similar acoustical and prosodic properties. In real situations, humans are helped by linguistic, contextual and other modalities like facial expressions, while interpreting the emotions from the speech.

For analyzing emotion specific prosodic characteristics, single male and female emotional utterances of IITKGP-SESC are considered. Means of the distribution of the prosodic features are computed and used for classifying eight emotions. Table 5.1 shows the mean values of the basic prosodic parameters of different emotions for both male and female speakers.

The mean duration is calculated by averaging the durations of all utterances. Mean pitch is computed by averaging the frame level pitch values for all utterances. Energy is an average of the frame level energies calculated for each utterance. Frames of size 20 ms and a shift of 10 ms are used for the above calculations. Though this statistical analysis of prosody toward emotion is very simple, it gives a clear insight of emotion specific knowledge present in the prosodic features. Table 5.2 shows the emotion recognition results based on the above prosodic parameters. Here a simple Euclidean distance measure is used to classify eight emotions. An average emotion recognition performance of around 45 and 51% is observed for male and female speech respectively. From the results, it may be observed that there are mis-classifications among high arousal emotions like anger,

Table 5.1   Mean values of the prosodic parameters for each of the emotions of IITKGP-SESC

	Male artist			Female artist		
Emotion	Duration (s)	Pitch (Hz)	Energy	Duration (s)	Pitch (Hz)	Energy
Anger	1.76	195.60	115.60	1.80	301.67	57.42
Disgust	1.62	188.05	73.42	1.67	308.62	54.01
Fear	1.79	210.70	147.31	1.89	312.07	76.65
Happiness	2.03	198.30	81.12	2.09	287.78	40.02
Neutral	1.93	184.37	83.13	2.04	267.13	40.89
Sadness	2.09	204.00	108.12	2.13	294.33	40.36
Sarcasm	2.16	188.44	98.57	2.20	301.11	34.20
Surprise	2.05	215.75	202.06	2.09	300.10	41.49

Table 5.2   Emotion classification performance using prosodic features

	Recognition performance (%) male, average: 45.38							
Emotions	Anger	Disgust	Fear	Happiness	Neutral	Sadness	Sarcasm	Surprise
Anger	37	0	10	23	13	0	0	17
Disgust	0	40	0	3	20	27	7	3
Fear	17	0	63	0	13	0	0	7
Happiness	33	0	10	37	10	0	0	10
Neutral	0	10	0	0	53	27	10	0
sadness	0	0	0	0	23	60	17	0
Sarcasm	0	27	0	0	10	20	43	0
Surprise	23	27	0	0	20	0	0	30
	Recognition performance (%) female, average: 50.88							
Anger	43	0	10	17	13	0	0	17
Disgust	0	47	0	0	3	0	27	23
Fear	13	0	67	10	3	0	0	7
Happiness	10	0	13	57	10	0	0	10
Neutral	0	17	0	0	50	23	0	10
sadness	0	7	0	0	33	60	0	0
Sarcasm	0	33	0	0	17	0	43	7
Surprise	3	37	10	0	0	10	0	40

happiness, and fear. Similar observations with respect to slow emotions such as disgust, sadness, and neutral are also seen. Most of the mis-classifications are biased toward neutral. Emotions expressed by female speakers are recognized fairly better than the emotions of male speakers.

From the first and second order statistics (mean and variance) of the prosodic parameters of one male and one female speaker, a qualitative analysis is presented in Table 5.3. From the symbols given in Table 5.3, it may be noted that for female speech, out of eight emotions, happiness, sadness, and sarcasm are with larger utterance level duration, whereas the duration for anger and disgust is less. Pitch values are very high for fear and disgust. Energy is observed to be very

Table 5.3 Generic trends of prosodic behavior of male and female utterances for eight different emotions

Emotion	Male artist						Female artist					
	Dur.	Ran.of dur.	Pitch	Ran.of pit.	Eng.	Ran.of eng.	Dur.	Ran.of dur.	Pitch	Ran.of pit.	Eng.	Ran.of eng.
Anger	<	>	=	<	=	=	<	>	>	=	=	=
Disgust	≪	<	<	<	≪	≪	≪	<	≫	<	<	<
Fear	=	<	>	=	≫	=	=	<	≫	≫	≫	≫
Happiness	>	<	=	=	<	=	>	=	=	=	<	=
Neutral	=	=	≪	=	<	=	=	≫	≪	<	<	≪
Sadness	>	=	=	=	>	≫	>	=	=	<	<	=
Sarcasm	≫	≫	<	≪	≪	<	≫	≪	>	>	≪	<
Surprise	>	≪	≫	≫	=	<	>	<	>	≫	<	<

Legend: ≪ – Very low, < – Low, = – Medium, > – High, ≫ – Very high *Dur.* Duration, *Ran.Dur.* Range of duration, *Ran.Pit.* Range of pitch, *Eng.* Energy, *Ran.Eng.* Range of energy

high for fear. It is interesting to note that these trends are not common between the genders. It indicates that the emotion expression cues, in the case of males and females, are slightly different. Table 5.3 represents the overall trend of the prosodic features related to the emotions expressed in the database IITKGP-SESC. The prosodic trends given in Table 5.3 are obtained based on the mean values of individual prosodic parameters computed over individual (local mean) and all emotions (global mean). A range with respect to overall mean (global mean) is fixed to qualitatively decide the trend of prosodic parameters for given emotions. Table 5.4 gives the details of the ranges of different prosodic parameters for deriving the trends. For example, to determine the trend of pitch with respect to anger, initially, the average pitch value of all the sentences of all the emotions (global mean) present in the database is computed. Later the average pitch of the anger sentences (local mean) is computed. The deviation of the pitch of anger sentences (local mean) from overall average pitch (global mean) helps to decide the trend of pitch for anger. Similarly, trends for duration and energy are determined by computing their global and local mean values. In Table 5.4, *LM* represents local mean (Mean of particular prosodic parameter for the specific emotion) and *GM* represents global mean (Mean of particular prosodic parameter for all the emotion).

5.3 Motivation

From Table 5.1, it may be observed that the average static prosodic values such as energy, pitch, and duration are distinguishable for different emotions. Similarly, the time varying dynamics in the prosodic contours also represent emotion specific information. Figure 5.1 shows the dynamics in prosodic contours for different emotions.

Table 5.4 Static ranges for qualitative description of prosodic parameters for different emotions, used in Table 5.3

Prosodic parameters	Very less	Less	Medium	High	Very high
Duration	$LM < 0.5GM$	$0.5GM \leq LM < 0.9GM$	$0.9GM \leq LM < 1.1GM$	$1.1GM \leq LM < 1.5GM$	$1.5GM < LM$
Pitch and energy	$LM < 0.9GM$	$0.9GM \leq LM < 0.95GM$	$0.95GM \leq LM < 1.05GM$	$1.05GM \leq LM < 1.1GM$	$1.1GM < LM$

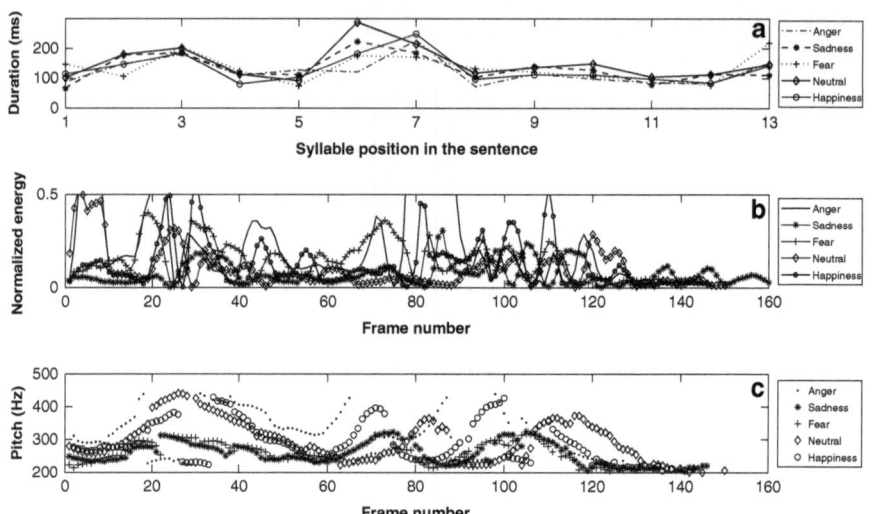

Fig. 5.1 (**a**) Duration patterns for the sequence of syllables, (**b**) energy contours and (**c**) pitch contours in different emotions for the utterance *mAtA aur pitA kA Adar karnA chAhie*

Obviously, there are inherent overlaps among these static and dynamic prosodic values with respect to the emotions. In the literature, several references are observed about using static prosodic features for speech emotion recognition. However, time dependent prosody variations may be used as the discrimination strategy, where static prosodic properties of different emotions show high overlap. Figure 5.1 shows three subplots indicating the (a) duration patterns of the sequence of syllables, (b) energy contours and (c) pitch contours of an utterance *mAtA aur pitA kA Adar karnA chAhie* in five different emotions. From the subplot indicating the duration patterns, one can observe the common trend of durations for all emotions. However, the trends also indicate that, for some emotions such as fear and happiness, the durations of the initial syllables of the utterance are longer, for happiness and neutral emotions middle syllables of the utterance seem to be longer, and the final syllables of the utterance seem to be longer for fear and anger (see Fig. 5.1a). From the energy plots, it is observed that the utterance with anger emotion has highest energy for the entire duration. Next to the anger emotion, fear and happiness show somewhat more energy than the other two emotions. The dynamics of energy contours can be used to discriminate fear and happiness (see Fig. 5.1b). It is observed from Fig. 5.1c that anger, happiness and neutral have somewhat higher pitch values, than the other two emotions. Using the dynamics (changes of prosodic values with respect to time) of pitch contours, easy discrimination is possible between anger, happiness and neutral emotions, even though they have similar average values. Thus, Fig. 5.1 provides the basic motivation to explore the dynamics of prosodic features for discriminating the emotions.

By observing pitch contours from Fig. 5.1c, it may be noted that for a given data, initial portions of the plots (the sequence of 20 pitch values) do not carry uniform discriminative information. Static features are almost the same for happiness and neutral. However, static features may be used to distinguish anger, sadness, fear emotions as their static pitch values are spread widely between 250 and 300 Hz. Similarly dynamic features are almost the same for all emotions except for fear. One may observe the initial decreasing and gradual rising trends of pitch contours for anger, happiness, neutral, and sadness emotions, whereas for the fear pitch contour starts with a rising trend. Similar local discriminative properties may also be observed in the case of energy and duration profiles from the initial, middle and final parts of the utterances. This phenomenon indicates that it may be sometimes difficult to classify the emotions based on either global or local prosodic trends derived from the entire utterance. Therefore, in this work, we intend to explore the static (global) and dynamic (local) prosodic features, along with their combination for speech emotion recognition at different levels (utterance, words, and syllables) and positions (initial, middle, and final).

5.4   Extraction of Global and Local Prosodic Features

In this chapter, emotion recognition (ER) systems are developed using local and global prosodic features, extracted from sentences. Sentence level static and dynamic prosodic features are derived by considering the entire sentence as a unit for feature extraction. Pitch contours are extracted using the zero frequency filter based method. The details of the zero frequency filter are given in Sect. 3.4.3 [130]. Zero frequency filter method determines the instants of significant excitation (epochs) present in the voiced regions of speech signal. Voiced regions are determined using frame level energy and periodicity. In the unvoiced region, the concept of pitch is not valid, hence pitch values are considered as zero for each interval of 10 ms. In the voiced region pitch is determined using epoch intervals. The time interval between successive epochs is known as the epoch interval. Reciprocal of the epoch interval is considered as the pitch at that instant of time. The energy contour of a speech signal is derived from the sequence of frame energies. Frame energies are computed by summing the squared sample amplitudes within a frame. Fourteen (two-duration, six-pitch, six-energy) prosodic parameters are identified to represent duration, pitch and energy components of global prosodic features. Average syllable and pause durations are considered as two duration parameters. Average syllable duration is computed as

$$ND_{syl} = \frac{D_s - D_p}{N_{syl}}$$

and average pause duration is computed as

$$ND_p = \frac{D_p}{D_s}$$

where ND_{syl} is average syllable duration, D_s is sentence duration, D_p is pause duration, N_{syl} is the number of syllables, and ND_p is average pause duration

Six pitch values are derived from the sentence level pitch contour. Those values are minimum, maximum, mean, standard deviation, median, and contour slope. The slopes of pitch and energy contours are determined by using the middle pitch and energy values of the first and the last words. These 14 values are concatenated in the order duration, pitch, and energy to form a feature vector that represents global prosody.

Local prosodic features are intended to capture the variations in the prosodic contours with respect to time. Therefore, the feature vector is expected to retain the natural dynamics of the prosody. In this regard, resampled energy and pitch contours are used to represent the feature vectors for local prosody. The dimension of pitch and energy contours is chosen to be 25, after evaluating the emotion recognition performance with 100, 50, 25, and 10 values. The recognition performance with 25 dimensional feature vectors is slightly better than the feature vectors with other dimensions. Here, the dimension 25 for pitch and energy contours is not crucial. The reduced size of the pitch and energy contours has to be chosen so that the dynamics of the original contours are retained in their resampled versions. The basic reasons for reducing the dimensionality of the original pitch and energy contours are (1) the need for the fixed dimensional input feature vectors for developing the SVM models and (2) the number of feature vectors required for training the classifier has to be proportional to the size of the feature vector to avoid the curse of dimensionality (The need of number of feature vectors grows exponentially as the dimensionality of feature vector increases. Therefore always there should be a proportion between the number of available feature vectors and their dimensionality). The local duration pattern is represented by the sequence of normalized syllable durations. Here the syllable durations are determined using the time interval between successive VOPs [156]. The length of duration contour is proportional to the number of syllables present in the sentence, which leads to feature vectors of unequal lengths. To obtain the feature vectors of equal length, the length of duration vector is fixed to be 18 (the maximum number of syllables present in the longest utterance of IITKGP-SESC). The length for shorter utterances is compensated by zero padding.

The English transcriptions of the text prompts of the Telugu database (IITKGP-SESC) are given in Table 5.5. The unicode set for Telugu alphabet is available at [157].

Out of ten speakers of IITKGP-SESC, the speech utterances of eight speakers (four male and four female) are used to train the emotion recognition models. Validation of the trained models is done using remaining two speakers' (one male

Table 5.5 English transcriptions of the Telugu text prompts of IITKGP-SESC

Sentence identity	Text prompts
S1	thallidhandrulanu gauravincha valenu
S2	mI kOsam chAlA sEpatnimchi chUsthunnAmu
S3	samAjamlo prathi okkaru chadhuvuko valenu
S4	ellappudu sathyamune paluka valenu
S5	I rOju nEnu tenali vellu chunnAnu
S6	kOpamunu vIdi sahanamunu pAtincha valenu
S7	anni dAnamulalo vidyA dAnamu minnA
S8	uchitha salahAlu ivvarAdhu
S9	dongathanamu cheyutA nEramu
S10	I rOju vAthAvaranamu podigA undhi
S11	dEsa vAsulandharu samaikhyAthA tho melaga valenu
S12	mana rAshtra rAjadhAni hyderAbAd
S13	sangha vidhrOha sekthulaku Ashrayam kalpincharAdhu
S14	thelupu rangu shAnthiki chihnamu
S15	gangA jalamu pavithra mainadhi

and one female) speech data. The details of the speech corpus, IITKGP-SESC, are given in Sect. 3.3 of Chap. 3. The description of development of emotion recognition models and their verification is discussed in the next section.

5.5 Results and Discussion

Emotion recognition systems are separately developed for sentence level global and local prosodic features. The combination of global and local level features is also explored to study emotion recognition (ER). In this work, we have considered eight emotions of IITKGP-SESC, for studying the role of global and local prosodic features in recognizing speech emotions. SVMs are used to develop emotion recognition models. Each SVM is trained with positive and negative examples. Positive feature vectors are derived from the utterances of the intended emotion, and negative feature vectors are derived from the utterances of all other emotions. Therefore, eight SVMs are developed to represent eight emotions. The basic block diagram of the ER system developed using SVMs is similar to the figure shown in Fig. 3.11 of Chap. 3. For evaluating the performance of the ER systems, the feature vectors are derived from the test utterances and are given as inputs to all eight trained emotion models. The output of each model is given to the decision module, where the category of the emotion is hypothesized based on the highest evidence among the eight emotion models.

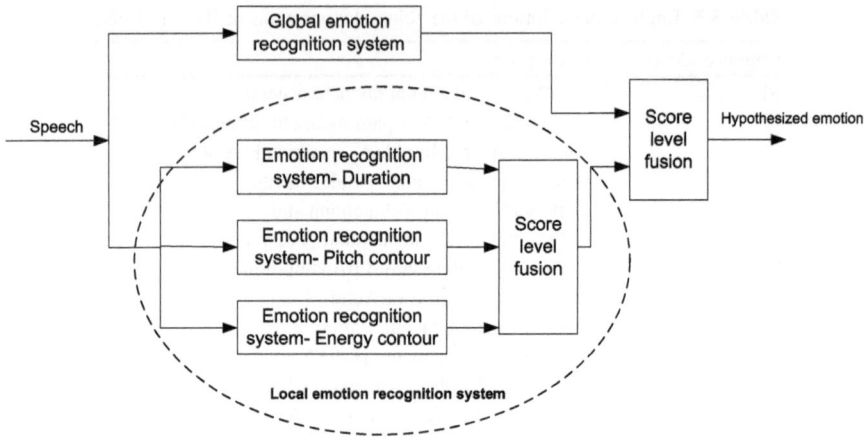

Fig. 5.2 Emotion recognition system using sentence level global and local prosodic features

Table 5.6   Emotion classification performance using global prosodic features computed over entire utterances. Average recognition performance: 43.75

| | Emotion recognition performance (%) | | | | | | | |
Emotions	Anger	Disgust	Fear	Happiness	Neutral	Sadness	sarcasm	Surprise
Anger	28	17	23	3	13	13	3	0
Disgust	7	47	0	0	3	10	33	0
Fear	7	0	67	7	0	10	0	9
Happiness	3	0	7	14	43	3	10	20
Neutral	0	0	7	17	67	0	3	6
Sadness	7	3	17	17	0	40	13	3
Sarcasm	0	10	0	13	20	3	44	10
Surprise	7	0	17	13	3	3	13	44

Abbreviations: *Ang.* Anger, *Dis.* Disgust, *Hap.* Happiness, *Neu.* Neutral, *Sar.* Sarcasm, *Sur.* Surprise

For analyzing the effect of global and local prosodic features on emotion recognition performance, separate models are developed using global and local prosodic features. The overall emotion recognition performance is obtained by combining the measures from the global and local prosodic features, as shown in Fig. 5.2.

The emotion recognition system based on global prosodic features consists of 8 emotion models, developed by using 14-dimensional feature vectors (duration parameters-2, pitch parameters-6, energy parameters-6). Emotion recognition performance of the models using global prosodic features is given in Table 5.6. Fear and neutral are recognized with the highest rate of 67%, whereas happiness utterances are identified with only 14% of accuracy. It is difficult to attain high performance, while classifying the underlying speech emotions using only static prosodic features. This is mainly due to the overlap of static prosodic features of different emotions. For instance, it is difficult to discriminate pairs like fear and anger, sarcasm and disgust using global prosodic features. Utterances of all eight emotions are mis-classified as either neutral, fear, or happiness. The mis-classification due to static prosodic features may be reduced by employing dynamic prosodic features for classification. Therefore, use of dynamic nature of prosody contours, captured through local prosodic features, is explored in this work for speech emotion recognition.

To study the relevance of individual local prosodic features in emotion recognition, three separate ER systems corresponding to sentence level duration, intonation and energy patterns are developed to capture local emotion specific information. Score level combination of these individual local prosodic systems is performed to obtain overall emotion recognition performance due to all local utterance level features. Emotion recognition performance using individual local prosodic features and their score level combination is given in Table 5.7.

The average emotion recognition performance due to individual local prosodic features is well above the performance of global prosodic features. The information of pitch dynamics has the highest discrimination of about 54%.

Energy and duration dynamic features have also achieved a recognition performance around 48%. From the results, it is observed that local prosodic features play a major role in discriminating the emotions. Score level combination of energy, pitch and duration features further improved the emotion recognition performance up to around 65%. Measures of emotion recognition models developed using global and local prosodic features are combined for improving the performance further. Table 5.8 shows the recognition performance of the emotion recognition system developed by combining the measures from global and local prosodic features. The average emotion recognition performance after combining the global and local prosodic features is observed to be about 65.63%. There is no considerable improvement in the emotion recognition rate, by combining the measures from global and local prosodic features. This indicates that the emotion discriminative properties of global prosodic feature are not complementary to those of local features. Therefore, local prosodic features alone would be sufficient to perform speech emotion recognition. The comparison of recognition performance in the case of each emotion, with respect to the global, local and their combination of features is shown in Fig. 5.3. It may be observed from the figure that anger, neutral, sadness, and surprise have achieved better discrimination using the combination of global and local prosodic features. Local prosodic features play an important role in the discrimination of disgust, happiness, and sarcasm. Fear is recognized well by using global prosodic features.

Table 5.7 Emotion classification performance using local prosodic features computed over entire sentences

| Emotions | Duration, average emotion recognition: 48.75 | | | | | | | |
	Anger	Disgust	Fear	Happiness	Neutral	Sadness	Sarcasm	Surprise
Anger	30	20	7	3	23	3	7	7
Disgust	7	67	3	0	10	3	10	0
Fear	7	7	53	0	10	10	7	6
Happiness	17	7	10	30	3	13	7	13
Neutral	7	3	3	3	57	21	3	3
Sadness	3	7	23	7	20	30	10	0
Sarcasm	0	13	4	10	0	0	73	0
Surprise	7	3	14	10	3	10	3	50
	Pitch, average emotion recognition: 53.75							
Anger	27	43	7	0	3	3	17	0
Disgust	10	60	10	0	0	3	7	0
Fear	3	13	43	7	0	10	7	17
Happiness	4	7	13	40	3	13	13	7
Neutral	3	7	0	3	80	7	0	0
Sadness	3	10	7	7	10	57	6	0
Sarcasm	0	0	7	3	0	10	63	17
Surprise	0	0	7	10	0	3	20	60
	Energy, average emotion recognition: 48							
Anger	43	37	7	0	7	0	6	0
Disgust	27	37	0	0	13	0	20	3
Fear	0	0	57	7	10	13	0	13
Happiness	7	7	10	43	17	7	7	2
Neutral	20	3	3	10	47	10	3	4
Sadness	0	3	13	17	17	40	10	0
Sarcasm	0	10	3	0	0	0	80	7
Surprise	0	0	37	13	3	0	10	37
	Duration + pitch + energy, average emotion recognition: 64.38							
Anger	40	4	23	27	3	0	3	0
Disgust	13	73	0	0	4	0	10	0
Fear	3	0	63	10	0	7	0	17
Happiness	7	0	10	57	3	13	3	7
Neutral	0	7	7	3	73	7	0	3
Sadness	0	7	13	7	0	63	10	0
Sarcasm	0	10	0	0	0	7	83	0
Surprise	3	0	17	10	0	0	7	63

Abbreviations: *Ang.* Anger, *Dis.* Disgust, *Hap.* Happiness, *Neu.* Neutral, *Sar.* Sarcasm, and *Sur.* Surprise

Table 5.8 Emotion classification performance using the combination of local and global prosodic features computed from entire utterances. Average recognition performance: 65.63%

Emotions 2–9	Emotion recognition performance (%)							
	Anger	Disgust	Fear	Happiness	Neutral	Sadness	Sarcasm	Surprise
Anger	47	40	3	0	3	0	7	0
Disgust	10	63	0	0	7	10	10	0
Fear	7	0	60	3	0	13	0	17
Happiness	10	0	7	53	20	0	3	7
Neutral	0	0	3	7	74	13	3	0
Sadness	0	0	17	3	0	77	3	0
Sarcasm	0	10	0	0	3	0	84	3
Surprise	0	0	10	17	0	0	6	67

Abbreviations: *Ang.* Anger, *Dis.* Disgust, *Hap.* Happiness, *Neu.* Neutral, *Sar.* Sarcasm, *Sur.* Surprise

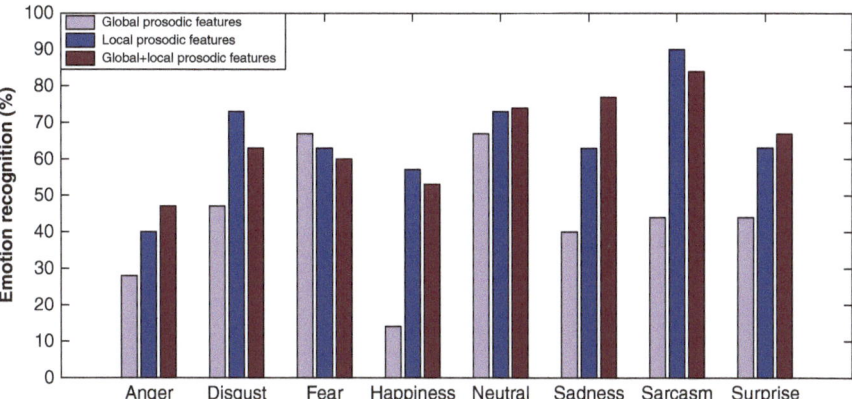

Fig. 5.3 Comparison of emotion recognition performance using utterance level global, local, and global + local prosodic features

5.6 Summary

In this chapter, prosodic analysis of speech signals has been performed at sentence level for the task of recognizing the underlying emotions. Eight emotions of IITKGP-SESC are used in this study. Support vector machines are used for developing the emotion models. Global and local prosodic features are separately extracted from utterances of speech for developing the emotion models. Global prosodic features are derived by computing statistical parameters like mean, maximum, minimum from the sequence of prosodic parameters. Local prosodic parameters are obtained from the sequence of syllable durations, frame level pitch and energy values. The prosodic contour trends are retained through local prosodic features. The combination of local and global prosodic features was found to marginally improve the performance compared to the performance of the systems developed using only local features.

Chapter 6
Summary and Conclusions

Abstract This chapter summarizes the research work presented in this book, highlights the contributions of the work and discusses the scope for future work.

The book is organized into six chapters. The first chapter introduces 'speech emotion recognition' as the contemporary research area. The second chapter critically reviews the research work done in the area of speech emotion recognition, with respect to different speech features for recognizing the underlying emotions [17]. At the end of this chapter, the motivation and scope for the present work are derived out of the literature review. In the third chapter, various excitation source features are proposed for discriminating the emotions. Spectral features derived from conventional block processing approach are proposed for speech emotion recognition in Chap. 4. Chapter 5 proposes use of static and dynamic prosodic features for recognition of emotions. Chapter 6 concludes the present work and flashes light on the directions for further research.

6.1 Summary of the Present Work

Recognizing emotions from speech has emerged as an important research area in the recent past. There are several applications where speech emotion recognition can be deployed. One of the important applications is developing an efficient speech interface to the machine. A properly designed emotional speech database is an essential component for carrying out research on speech emotion recognition. In this work, the simulated emotional speech database in Telugu language (second highest spoken language in India) is collected using radio artists in eight common emotions. This database is named as the Indian Institute of Technology KharaGPur-Simulated Emotion Speech Corpus (IITKGP-SESC). The emotions present in the database are anger, disgust, fear, happiness, neutral, sadness, sarcasm, and surprise.

S.R. Krothapalli and S.G. Koolagudi, *Emotion Recognition using Speech Features*, SpringerBriefs in Electrical and Computer Engineering, DOI 10.1007/978-1-4614-5143-3_6, © Springer Science+Business Media New York 2013

The recognition of speech emotions using different speech features is studied using IITKGP-SESC. In this book, three information sources namely (1) Excitation source, (2) Vocal tract system, and (3) Prosody are explored for recognition of speech emotions.

From the literature, it is observed that excitation source features are completely ignored from use for any of the speech tasks, by considering it as the unpredictable part of the speech. There is no authentic evidence for using excitation source information, especially for the task of speech emotion recognition. With this notion, different excitation source features are proposed for capturing emotion specific information from the speech utterances. LP residual signal, LP residual phase, epoch parameters, GVV signal and GVV parameters are proposed as excitation source features for recognizing the speech emotions. For capturing the emotion specific information from the higher order relations present in the samples of the LP residual, the phase of the LP residual samples and the GVV signal, AANN models are explored. SVMs are used for discriminating the emotions from the utterance level epoch parameters. All the proposed excitation source features have demonstrated the presence of emotion specific information by discriminating the emotions with around 50% accuracy. Excitation source features extracted from glottal closure regions, and epoch parameters at the utterance-level seem to carry relatively more emotion discriminative information than the other excitation source features.

Most of the emotion recognition studies, based on spectral features, have employed conventional block processing, where speech features are derived from the entire speech signal. In this work also the basic system features such as LPCCs, MFCCs and formant features are used to characterize the emotions. High amplitude regions of the spectrum are robust while developing speech systems using noisy or real time speech data. In this book, we have used formant features along with other spectral features for recognizing the emotions. In this work GMMs are used for developing the emotion models using spectral features. Among different spectral features, LPCCs seem to perform better in view of discriminating emotions. Recognition performance using formant features is not appreciable, but formant features in combination with other features have shown improvement in the performance.

Prosodic features are treated as the effective correlates of speech emotions. In the literature, static prosodic features have been thoroughly investigated for emotion recognition. However, from the perceptual observation of emotional speech, it is observed that emotions are gradually manifested through the sequence of phonemes. This gradual manifestation of emotions may be captured through variations in the articulator movements, while producing emotions. With this motivation, in this book, temporal variations of prosody contours are proposed to capture the emotion specific information. Global and local prosodic features extracted from sentences are explored to discriminate the emotions. In this work, SVM models are explored for capturing the emotion discriminative information from the local and global prosodic features. From the recognition studies using the proposed prosodic features, it is observed that local prosodic features that represent the temporal variation in prosody have more discriminative ability, while classifying the emotions.

6.2 Contributions of the Present Work

Some of the primary contributions of this book are:

- Design and development of an emotional speech database in Telugu to promote research on speech emotion processing in an Indian context. Design and development of a Hindi movie database to represent real life-like emotions for modeling natural emotions.
- Different excitation source features are proposed and they are systematically analyzed for speech emotion recognition.
- Utterance wise spectral features extracted form a speech segment of 20 ms, are proposed for recognizing the speech emotions.
- Local prosodic features indicating temporal variations in prosodic contours are proposed for recognizing speech emotions.

6.3 Conclusions from the Present Work

- Though excitation source information appears like random noise and considered not to contain any information beyond the fundamental frequency, due to the presence of higher order correlations, it contains useful emotion discriminative information. The proposed excitation source features may not be sufficient to develop sophisticated emotion recognition systems. However, the proposed source features may provide supplementary measures to other features while discriminating the emotions.
- Spectral features extracted from an entire speech signal are proved to capture useful emotion specific information as they did in the cases of other speech systems. Out of different spectral features proposed, LPCCs perform better in modeling the emotions.
- Temporal variations in prosodic contours represented by local prosodic features provide more emotion discriminative information than the global prosodic features.
- Two stage emotion classification with broad classification of emotions based on speaking rate at the first stage and finer classification of broad group emotions at the second level has improved the performance compared to single stage emotion classification.

6.4 Scope for Future Work

- The majority of the research results produced on emotion speech recognition have used databases with limited numbers of speakers. The work discussed in this book also reported the performance based on ten speakers' speech corpora (IITKGP-SESC and Emo-DB). While developing emotion recognition systems using limited speaker databases, speaker specific information may play a considerable role, if speech utterances of the same speakers are used for training and testing the models. On the other hand, developed models may produce poor results, due to lack of generality, if speech utterances of different speakers are used for training and testing the models. Therefore, there is a need of larger emotional speech databases with reasonably large numbers of speakers and text prompts. Emotion recognition studies have to be conducted on large databases in view of speaker, text and session variabilities.

- This book mainly focuses on characterizing the emotions from a classification point of view. Hence, the main task carried out was deriving the emotion specific information from speech, and using it for classifying the emotions. On the other hand, emotion synthesis through speech is also an important task. Here, emotion specific information may be predicted from the text, and then it has to be incorporated during synthesis. For predicting the emotion specific information, appropriate models have to be developed using a sufficiently large emotion speech corpus [158]. In emotion synthesis, the major issues are design of accurate prediction models and preparation of an appropriate emotion speech corpus.

- Expression of emotions is an universal phenomenon, which may be independent of speaker, gender and language. Cross-lingual emotion recognition study may be another interesting work for further research. The emotion recognition models developed using the utterances of a particular language should yield appreciably good recognition performance for any test utterance of the other language.

- The majority of the work done and results produced in this work are on recognizing speech emotions using a simulated Telugu database. The real challenge is to recognize speech emotions from natural emotions. The features and techniques discussed in this book may be applied to a natural speech corpus, to analyze emotion recognition. Realization of this needs the collection of a good natural emotional speech corpus, covering a wide range of emotions, which is another challenge.

- In this work, conventional spectral features are explored for discriminating the emotions. These features are extracted from block processing approach, hence they represent average spectral characteristics over a frame. For capturing the finer variations in spectrum, spectral features extracted from each pitch cycle may be useful [152]. These finer variations in spectral characteristics may contain emotion discriminative information. In addition to examining each pitch cycle, glottal closure and open phases may be separately examined for deriving the complementary or supplementary emotion specific information [159, 160].

- In this study global and local prosodic features are examined at the sentence level for the recognition of emotions. One can extend the present framework to examine the global and local prosodic features extracted from word and syllable levels for discriminate the emotions [161]. Positional information related to words and syllables may be explored for resolving the ambiguity of confusable emotions [162].

- In this work, emotion recognition was analysed independently using excitation source features, vocal tract system features and prosodic features. The recognition performance may be further enhanced by exploring the appropriate fusion techniques at feature and model levels.
- In this work, a mostly emotion classification task is performed using a single model (i.e., GMM, AANN, or SVM). In future work, hybrid models can be explored for improving the recognition performance. The basic idea behind using the hybrid models is that they derive the evidence from different perspectives, and hence, the combination of evidence may enhance the performance, if the evidences are supplementary in nature.
- The proposed features and methods in this book are evaluated using a simulated Telugu emotional database and the obtained results are verified on an internationally known Berlin emotion speech database. The trend of emotion recognition is not known on other Indian languages. It would be nice to evaluate the proposed features on different Indian languages for emotion recognition. This helps to comment on whether the methods and features used in this work are language independent. This analysis is also helpful to group the languages based on their emotion characteristics, which in turn would improve the performance of language identification systems.
- In this book, speech emotion recognition is analyzed using different speech features. The study on discrimination of emotions may be extended to the emotion dimensions (arousal, valance and power), which are derived from the psychology of production and perception of emotions. Deriving the appropriate speech features related to the emotion dimensions can be explored for further improving the recognition performance.
- Expression of emotions is a multi-modal activity. Therefore, other modalities like facial expression and bio-signals may be used as the supportive evidence along with the speech signal for developing the robust emotion recognition systems.
- The effect of emotion expression also depends upon the linguistic contents of the speech. Identification of emotion-salient words from the emotional speech, and the features extracted from these words along with other conventional features may enhance the emotion recognition performance.
- In real-time applications such as call analysis in emergency services like ambulance and fire brigade, verification of emotions to analyze the genuineness of the requests is important. In this context, under the framework of emotion verification appropriate features and models can be explored.
- While extracting emotion specific information from the epoch parameters, the accuracy of epoch locations play an important role. The existing epoch extraction methods perform well in the case of clean speech. Therefore, there is a need to strengthen the existing epoch extraction methods to suit real world situations of speech recording, where recorded speech may contain different background environments. Application of an improved epoch extraction method may further improve the emotion recognition performance.

- While computing formant features, a simple peak picking algorithm is used to pick up the four dominant peaks from the spectrum. It is observed that sometimes spurious peaks are detected as the formants. Using an efficient formant detection technique may improve emotion recognition performance further.
- Most of today's emotion recognition systems experience a high influence of speaker specific information during emotion classification. In the present work, we normalized the features to minimize speaker dependent information. This is not proved to be an efficient method to nullify speaker specific information. An efficient technique may be developed to remove speaker specific information from the speech utterance.

Appendix A
Linear Prediction Analysis of Speech

Linear prediction (LP) is one of the most important tools in speech analysis. The philosophy behind linear prediction is that a speech sample can be approximated as a linear combination of past samples. Then, by minimizing the sum of the squared differences between the actual speech samples and the linearly predicted ones over a finite interval, a unique set of predictor coefficients can be determined [15]. LP analysis decomposes the speech into two highly independent components, the vocal tract parameters (LP coefficients) and the glottal excitation (LP residual). It is assumed that speech is produced by exciting a linear time-varying filter (the vocal tract) by random noise for unvoiced speech segments, or a train of pulses for voiced speech. Figure A.1 shows a model of speech production for LP analysis [163]. It consists of a time varying filter $H(z)$ which is excited by either a quasi periodic or a random noise source.

The most general predictor form in linear prediction is the autoregressive moving average (ARMA) model where the speech sample $s(n)$ is modelled as a linear combination of the past outputs and the present and past inputs [164–166]. It can be written mathematically as follows

$$s(n) = -\sum_{k=1}^{p} a_k s(n-k) + G \sum_{l=0}^{q} b_l u(n-l), \quad b_0 = 1 \tag{A.1}$$

where $a_k, 1 \le k \le p, b_l, 1 \le l \le q$ and gain G are the parameters of the filter. Equivalently, in frequency domain, the transfer function of the linear prediction speech model is [164]

$$H(z) = \frac{1 + \sum_{l=1}^{q} b_l z^{-l}}{1 + \sum_{k=1}^{p} a_k z^{-k}}. \tag{A.2}$$

$H(z)$ is referred to as a pole-zero model. The zeros represent the nasals and the poles represent the resonances (formants) of the vocal-tract. When $a_k = 0$ for $1 \le k \le p$,

S.R. Krothapalli and S.G. Koolagudi, *Emotion Recognition using Speech Features*,
SpringerBriefs in Electrical and Computer Engineering, DOI 10.1007/978-1-4614-5143-3,
© Springer Science+Business Media New York 2013

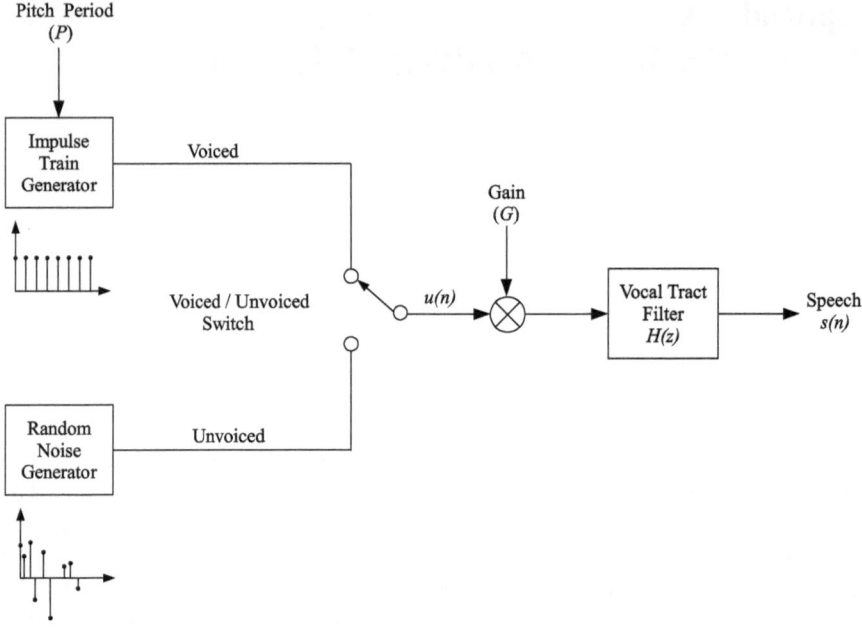

Fig. A.1 Model of speech production for LP analysis

$H(z)$ becomes an all-zero or moving average (MA) model. Conversely, when $b_l = 0$ for $1 \leq l \leq q$, $H(z)$ becomes an all-pole or autoregressive (AR) model [165]. For non-nasal voiced speech sounds the transfer function of the vocal-tract has no zeros whereas the nasals and unvoiced sounds usually includes the poles (resonances) as well as zeros (anti resonances) [163].

Generally the all-pole model is preferred for most applications because it is computationally more efficient and its the acoustic tube model for speech production. It can model sounds such as vowels well enough. The zeros arise only in nasals and in unvoiced sounds like fricatives. These zeros are approximately modelled by including more poles [164]. In addition, the location of a poles considerably more important perceptually than the location of a zero [167]. Moreover, it is easy to solve an all-pole model. To solve a pole-zero model, it is necessary to solve a set of nonlinear equations, but in the case of an all-pole model, only a set of linear equations need to be solved. The transfer function of the all-pole model is [165]

$$H(z) = \frac{G}{1 + \sum\limits_{k=1}^{p} a_k z^{-k}}. \tag{A.3}$$

The number p implies that the past p output samples are being considered, which is also the order of the linear prediction. With this transfer function, we get a difference equation for synthesizing the speech samples $s(n)$ as

$$s(n) = -\sum_{k=1}^{p} a_k s(n-k) + Gu(n) \tag{A.4}$$

where the coefficients a_k's are known as linear predictive coefficients (LPCs) and p is the order of the LP filter. It should be selected such that there are at least a pair of poles per each formant. Generally, the prediction order is chosen using the relation

$$p = 2 \times (BW + 1) \tag{A.5}$$

where BW is the speech bandwidth in kHz.

A.1   The Prediction Error Signal

The error signal or the residual signal $e(n)$ is the difference between the input speech and the estimated speech [165].

$$e(n) = s(n) + \sum_{k=1}^{p} a_k s(n-k). \tag{A.6}$$

Here the gain G is usually ignored to allow the parameterizations to be independent of the signal intensity. In z-domain $e(n)$ can be viewed as the output of the prediction filter $A(z)$ to the input speech signal $s(n)$ which is expressed as

$$E(z) = A(z)S(z) \tag{A.7}$$

where

$$A(z) = \frac{1}{H(z)} = 1 + \sum_{k=1}^{p} a_k z^{-k}; \quad G = 1. \tag{A.8}$$

The LP residual represents the excitations for production of speech [168]. The residual is typically a series of pulses, when derived from voiced speech or noise-like, when derived from unvoiced speech. The whole LP model can be decomposed into the two parts, the analysis part and the synthesis part as shown in Fig. A.2. The LP analysis filter removes the formant structure of the speech signal and leaves a lower energy output prediction error which is often called the LP residual or excitation signal. The synthesis part takes the error signal as an input [165]. The input is filtered by the synthesis filter $1/A(z)$, and the output is the speech signal.

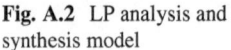

Fig. A.2 LP analysis and
synthesis model

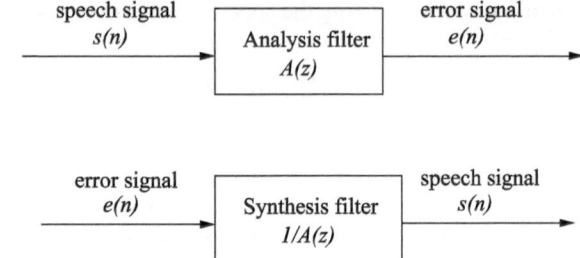

A.2   Estimation of Linear Prediction Coefficients

There are two widely used methods for estimating the LP coefficients (LPCs):
(i) Autocorrelation and (ii) Covariance. Both methods choose the short term filter
coefficients (LPCs) a_k in such a way that the energy in the error signal (residual)
is minimized. For speech processing tasks, the autocorrelation method is almost
exclusively used because of its computational efficiency and inherent stability
whereas the covariance method does not guarantee the stability of the all-pole
LP synthesis filter [163, 169]. The autocorrelation method of computing LPCs is
described below:

First, speech signal $s(n)$ is multiplied by a window $w(n)$ to get the windowed
speech segment $s_w(n)$. Normally, a Hamming or Hanning window is used. The
windowed speech signal is expressed as

$$s_w(n) = s(n)w(n). \tag{A.9}$$

The next step is to minimize the energy in the residual signal. The residual energy
E_p is defined as [165]

$$E_p = \sum_{n=-\infty}^{\infty} e^2(n) = \sum_{n=-\infty}^{\infty} \left(s_w(n) + \sum_{k=1}^{p} a_k s_w(n-k) \right)^2. \tag{A.10}$$

The values of a_k that minimize E_p are found by setting the partial derivatives of the
energy E_p with respect to the LPC parameters equal to 0.

$$\frac{\partial E_p}{\partial a_k} = 0, \quad 1 \le k \le p. \tag{A.11}$$

This results in the following p linear equations for the p unknown parameters
$a_1, \dots, a_p$

$$\sum_{k=1}^{p} a_k \sum_{n=-\infty}^{\infty} s_w(n-i)s_w(n-k) = -\sum_{n=-\infty}^{\infty} s_w(n-i)s_w(n), \quad 1 \le i \le p. \tag{A.12}$$

This linear equations can be expressed in terms of the autocorrelation function. This is because the autocorrelation function of the windowed segment $s_w(n)$ is defined as

$$R_s(i) = \sum_{n=-\infty}^{\infty} s_w(n)s_w(n+i), \qquad 1 \le i \le p. \tag{A.13}$$

Exploiting the fact that the autocorrelation function is an even function i.e., $R_s(i) = R_s(-i)$. By substituting the values from Eq. (A.13) in Eq. (A.12), we get

$$\sum_{k=1}^{p} R_s(|i-k|)a_k = -R_s(i), \qquad 1 \le i \le p. \tag{A.14}$$

These set of p linear equations can be represented in the following matrix form as [163]

$$\begin{bmatrix} R_s(0) & R_s(1) & \cdots & R_s(p-1) \\ R_s(1) & R_s(0) & \cdots & R_s(p-2) \\ \vdots & \vdots & \ddots & \vdots \\ R_s(p-1) & R_s(p-2) & \cdots & R_s(0) \end{bmatrix} \begin{bmatrix} a_1 \\ a_2 \\ \vdots \\ a_p \end{bmatrix} = - \begin{bmatrix} R_s(1) \\ R_s(2) \\ \vdots \\ R_s(p) \end{bmatrix} \tag{A.15}$$

This can be summarized using vector-matrix notation as

$$\mathbf{R_s a} = -\mathbf{r_s} \tag{A.16}$$

where the $p \times p$ matrix $\mathbf{R_s}$ is known as the autocorrelation matrix. The resulting matrix is a Toeplitz matrix where all elements along a given diagonal are equal. This allows the linear equations to be solved by the Levinson-Durbin algorithm. Because of the Toeplitz structure of $\mathbf{R_s}$, $A(z)$ is minimum phase [163]. At the synthesis filter $H(z) = 1/A(z)$, the zeros of $A(z)$ become the poles of $H(z)$. Thus, the minimum phase of $A(z)$ guarantees the stability of $H(z)$.

Appendix B
MFCC Features

The MFCC feature extraction technique basically includes windowing the signal, applying the DFT, taking the log of the magnitude and then warping the frequencies on a Mel scale, followed by applying the inverse DCT. The detailed description of various steps involved in the MFCC feature extraction is explained below.

1. **Pre-emphasis:** Pre-emphasis refers to filtering that emphasizes the higher frequencies. Its purpose is to balance the spectrum of voiced sounds that have a steep roll-off in the high frequency region. For voiced sounds, the glottal source has an approximately -12 dB/octave slope [170]. However, when the acoustic energy radiates from the lips, this causes a roughly $+6$ dB/octave boost to the spectrum. As a result, a speech signal when recorded with a microphone from a distance has approximately a -6 dB/octave slope downward compared to the true spectrum of the vocal tract. Therefore, pre-emphasis removes some of the glottal effects from the vocal tract parameters. The most commonly used pre-emphasis filter is given by the following transfer function

$$H(z) = 1 - bz^{-1} \tag{B.1}$$

   where the value of b controls the slope of the filter and is usually between 0.4 and 1.0 [170].

2. **Frame blocking and windowing:** The speech signal is a slowly time-varying or quasi-stationary signal. For stable acoustic characteristics, speech needs to be examined over a sufficiently short period of time. Therefore, speech analysis must always be carried out on short segments across which the speech signal is assumed to be stationary. Short-term spectral measurements are typically carried out over 20 ms windows, and advanced every 10 ms [171, 172]. Advancing the time window every 10 ms enables the temporal characteristics of individual speech sounds to be tracked and the 20 ms analysis window is usually sufficient to provide good spectral resolution of these sounds, and at the same time short enough to resolve significant temporal characteristics. The purpose of the overlapping analysis is that each speech sound of the input sequence would be

S.R. Krothapalli and S.G. Koolagudi, *Emotion Recognition using Speech Features*, 105
SpringerBriefs in Electrical and Computer Engineering, DOI 10.1007/978-1-4614-5143-3,
© Springer Science+Business Media New York 2013

approximately centered at some frame. On each frame a window is applied to taper the signal towards the frame boundaries. Generally, Hanning or Hamming windows are used [170]. This is done to enhance the harmonics, smooth the edges and to reduce the edge effect while taking the DFT on the signal.

3. **DFT spectrum:** Each windowed frame is converted into magnitude spectrum by applying DFT.

$$X(k) = \sum_{n=0}^{N-1} x(n)e^{\frac{-j2\pi nk}{N}}; \quad 0 \leq k \leq N-1 \tag{B.2}$$

where N is the number of points used to compute the DFT.

4. **Mel-spectrum:** Mel-Spectrum is computed by passing the Fourier transformed signal through a set of band-pass filters known as mel-filter bank. A mel is a unit of measure based on the human ears perceived frequency. It does not correspond linearly to the physical frequency of the tone, as the human auditory system apparently does not perceive pitch linearly. The mel scale is approximately a linear frequency spacing below 1 kHz, and a logarithmic spacing above 1 kHz [173]. The approximation of mel from physical frequency can be expressed as

$$f_{mel} = 2,595 \log_{10}\left(1 + \frac{f}{700}\right) \tag{B.3}$$

where f denotes the physical frequency in Hz, and f_{mel} denotes the perceived frequency [171].

Filter banks can be implemented in both time domain and frequency domain. For MFCC computation, filter banks are generally implemented in frequency domain. The center frequencies of the filters are normally evenly spaced on the frequency axis. However, in order to mimic the human ears perception, the warped axis according to the non-linear function given in Eq. (B.3), is implemented. The most commonly used filter shaper is triangular, and in some cases the Hanning filter can be found [170]. The triangular filter banks with mel-frequency warping is given in Fig. B.1.

The mel spectrum of the magnitude spectrum $X(k)$ is computed by multiplying the magnitude spectrum by each of the of the triangular mel weighting filters.

$$s(m) = \sum_{k=0}^{N-1} \left[|X(k)|^2 H_m(k)\right]; \quad 0 \leq m \leq M-1 \tag{B.4}$$

where M is total number of triangular mel weighting filters [174, 175]. $H_m(k)$ is the weight given to the kth energy spectrum bin contributing to the mth output band and is expressed as:

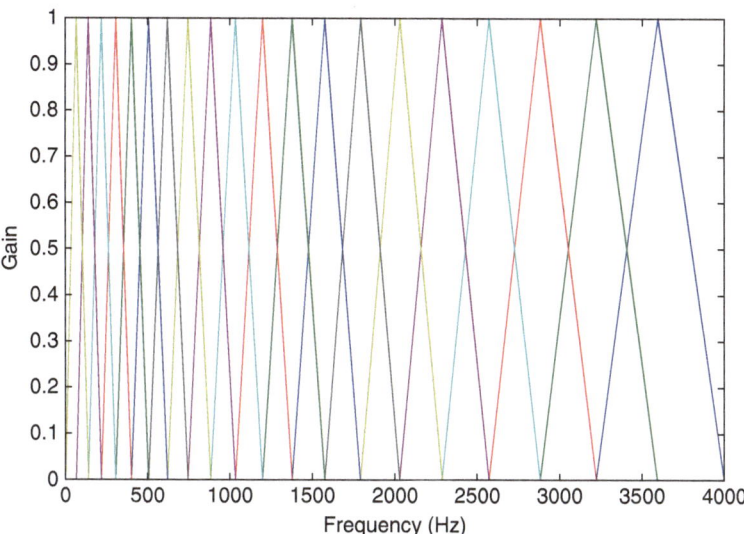

Fig. B.1 Mel-filter bank

$$H_m(k) = \begin{cases} 0, & k < f(m-1) \\ \frac{2(k-f(m-1))}{f(m)-f(m-1)}, & f(m-1) \le k \le f(m) \\ \frac{2(f(m+1)-k)}{f(m+1)-f(m)}, & f(m) < k \le f(m+1) \\ 0, & k > f(m+1) \end{cases} \tag{B.5}$$

with m ranging from 0 to $M-1$.

5. **Discrete Cosine Transform (DCT):** Since the vocal tract is smooth, the energy levels in adjacent bands tend to be correlated. The DCT is applied to the transformed mel frequency coefficients produces a set of cepstral coefficients. Prior to computing DCT the mel spectrum is usually represented on a log scale. This results in a signal in the cepstral domain with a que-frequency peak corresponding to the pitch of the signal and a number of formants representing low que-frequency peaks. Since most of the signal information is represented by the first few MFCC coefficients, the system can be made robust by extracting only those coefficients ignoring or truncating higher order DCT components [170]. Finally, MFCC is calculated as [170]

$$c(n) = \sum_{m=0}^{M-1} \log_{10}(s(m)) \cos\left(\frac{\pi n(m-0.5)}{M}\right); \qquad n = 0, 1, 2, \ldots, C-1 \tag{B.6}$$

where $c(n)$ are the cepstral coefficients and C is the number of MFCCs. Traditional MFCC systems use only 8–13 cepstral coefficients. The zeroth coefficient is often excluded since it represents the average log-energy of the input signal, which only carries little speaker-specific information.

6. **Dynamic MFCC features:** The cepstral coefficients are usually referred to as static features, since they only contain information from a given frame. The extra information about the temporal dynamics of the signal is obtained by computing first and second derivatives of cepstral coefficients [176, 177]. The first order derivative is called delta coefficients, and the second order derivative is called delta-delta coefficients. Delta coefficients tell about the speech rate, and delta-delta coefficients provide information similar to acceleration of speech. The commonly used definition for computing dynamic parameter is

$$\Delta c_m(n) = \frac{\sum\limits_{i=-T}^{T} k_i c_m(n+i)}{\sum\limits_{i=-T}^{T} |i|} \tag{B.7}$$

where $c_m(n)$ denotes the mth feature for the nth time frame, k_i is the ith weight and T is the number of successive frames used for computation. Generally T is taken as 2. The delta-delta coefficients are computed by taking the first order derivative of the delta coefficients.

Appendix C
Gaussian Mixture Model (GMM)

In the speech and speaker recognition the acoustic events are usually modeled by Gaussian probability density functions (PDFs), described by the mean vector and the covariance matrix. However unimodel PDF with only one mean and covariance are unsuitable to model all variations of a single event in speech signals. Therefore, a mixture of single densities is used to model the complex structure of the density probability. For a D-dimensional feature vector denoted as x_t, the mixture density for speaker Ω is defined as weighted sum of M component Gaussian densities as given by the following [178]

$$P(x_t|\Omega) = \sum_{i=1}^{M} w_i P_i(x_t) \tag{C.1}$$

where w_i are the weights and $P_i(x_t)$ are the component densities. Each component density is a D-variate Gaussian function of the form

$$P_i(x_t) = \frac{1}{(2\pi)^{D/2} |\Sigma_i|^{\frac{1}{2}}} e^{-\frac{1}{2}[(x_t-\mu_i)'\Sigma_i^{-1}(x_t-\mu_i)]} \tag{C.2}$$

where μ_i is a mean vector and Σ_i covariance matrix for ith component. The mixture weights have to satisfy the constraint [178]

$$\sum_{i=1}^{M} w_i = 1. \tag{C.3}$$

The complete Gaussian mixture density is parameterized by the mean vector, the covariance matrix and the mixture weight from all component densities. These parameters are collectively represented by

$$\Omega = \{w_i, \mu_i, \Sigma_i\}; \qquad i = 1, 2, \ldots . M. \tag{C.4}$$

S.R. Krothapalli and S.G. Koolagudi, *Emotion Recognition using Speech Features*, SpringerBriefs in Electrical and Computer Engineering, DOI 10.1007/978-1-4614-5143-3, © Springer Science+Business Media New York 2013

C.1   Training the GMMs

To determine the model parameters of GMM of the speaker, the GMM has to be trained. In the training process, the maximum likelihood (ML) procedure is adopted to estimate model parameters. For a sequence of training vectors $X = \{x_1, x_2, \ldots, x_T\}$, the GMM likelihood can be written as (assuming observations independence) [178]

$$P(X|\Omega) = \prod_{t=1}^{T} P(x_t|\Omega). \qquad (C.5)$$

Usually this is done by taking the logarithm and is commonly named as log-likelihood function. From Eqs. (C.1) to (C.5), the log-likelihood function can be written as

$$\log[P(X|\Omega)] = \sum_{t=1}^{T} \log \left[\sum_{i=1}^{M} w_i P_i(x_t) \right]. \qquad (C.6)$$

Often, the average log-likelihood value is used by dividing $\log[P(X|\Omega)]$ by T. This is done to normalize out duration effects from the log-likelihood value. Also, since the incorrect assumption of independence is underestimating the actual likelihood value with dependencies, scaling by T can be considered a rough compensation factor [179]. The parameters of a GMM model can be estimated using maximum likelihood (ML) estimation. The main objective of the ML estimation is to derive the optimum model parameters that can maximize the likelihood of GMM. The likelihood value is, however, a highly nonlinear function in the model parameters and direct maximization is not possible. Instead, maximization is done through iterative procedures. Of the many techniques developed to maximize the likelihood value, the most popular is the iterative expectation maximization (EM) algorithm [180].

C.1.1   Expectation Maximization (EM) Algorithm

The EM algorithm begins with an initial model Ω and tends to estimate a new model such that the likelihood of the model increasing with each iteration. This new model is considered to be an initial model in the next iteration and the entire process is repeated until a certain convergence threshold is obtained or a certain predetermined number of iterations have been made. A summary of the various steps followed in the EM algorithm are described below.

1. **Initialization:** In this step an initial estimate of the parameters is obtained. The performance of the EM algorithm depends on this initialization. Generally, LBG [181] or K-means algorithm [182, 183] is used to initialize the GMM parameters.

2. **Likelihood Computation:** In each iteration the posterior probabilities for the ith mixture is computed as [178]:

$$\Pr(i|x_t) = \frac{w_i P_i(x_t)}{\sum\limits_{j=1}^{M} w_j P_j(x_t)}. \tag{C.7}$$

3. **Parameter Update:** Having the posterior probabilities, the model parameters are updated according to the following expressions [178].

    Mixture weight update:

$$\overline{w_i} = \frac{\sum\limits_{i=1}^{T} \Pr(i|x_t)}{T}. \tag{C.8}$$

    Mean vector update:

$$\overline{\mu_i} = \frac{\sum\limits_{i=1}^{T} \Pr(i|x_t) x_t}{\sum\limits_{i=1}^{T} \Pr(i|x_t)}. \tag{C.9}$$

    Covariance matrix update:

$$\overline{\sigma_i^2} = \frac{\sum\limits_{i=1}^{T} \Pr(i|x_t)\,|x_t - \overline{\mu_i}|^2}{\sum\limits_{i=1}^{T} \Pr(i|x_t)}. \tag{C.10}$$

In the estimation of the model parameters, it is possible to choose, either full covariance matrices or diagonal covariance matrices. It is more common to use diagonal covariance matrices for GMM, since linear combination of diagonal covariance Gausses has the same model capability with full matrices [184]. Another reason is that speech utterances are usually parameterized with cepstral features. Cepstral features are more compactible, discriminative, and most important, they are nearly uncorrelated, which allows diagonal covariance to be used by the GMMs [178,185]. The iterative process is normally carried out ten times, at which point the model is assumed to converge to a local maximum [178].

C.1.2 Maximum A Posteriori (MAP) Adaptation

Gaussian mixture models for a speaker can be trained using the modeling described earlier. For this, it is necessary that sufficient training data is available in order to create a model of the speaker. Another way of estimating a statistical model, which is especially useful when the training data available is of short duration, is by using maximum a posteriori adaptation (MAP) of a background model trained on the speech data of several other speakers [186]. This background model is a large GMM that is trained with a large amount of data which encompasses the different kinds of speech that may be encountered by the system during training. These different kinds may include different channel conditions, composition of speakers, acoustic conditions, etc. A summary of MAP adaptation steps are given below.

For each mixture i from the background model, $Pr(i|x_t)$ is calculated as [187]

$$Pr(i|x_t) = \frac{w_i P_i(x_t)}{\sum\limits_{j=1}^{M} w_j P_j(x_t)}. \tag{C.11}$$

Using $Pr(i|x_t)$, the statistics of the weight, mean and variance are calculated as follows [187]

$$n_i = \sum_{i=1}^{T} Pr(i|x_t) \tag{C.12}$$

$$E_i(x_t) = \frac{\sum\limits_{i=1}^{T} Pr(i|x_t)x_t}{n_i} \tag{C.13}$$

$$E_i(x_t^2) = \frac{\sum\limits_{i=1}^{T} Pr(i|x_t)x_t^2}{n_i}. \tag{C.14}$$

These new statistics calculated from the training data are then used adapt the background model, and the new weights ($\hat{w}_i$), means ($\hat{\mu}_i$) and variances ($\hat{\sigma}_i^2$) are given by Reynolds [187]

$$\hat{w}_i = \left[\frac{\alpha_i n_i}{T} + (1 - \alpha_i)w_i \right] \gamma \tag{C.15}$$

$$\hat{\mu}_i = \alpha_i E_i(x_t) + (1 - \alpha_i)\mu_i \tag{C.16}$$

$$\hat{\sigma}_i^2 = \alpha_i E_i(x_t^2) + (1 - \alpha_i)(\sigma_i^2 + \mu_i^2) - \hat{\mu}_i^2. \tag{C.17}$$

A scale factor γ is used, which ensures that all the new mixture weights sum to 1. α_i is the adaptation coefficient which controls the balance between the old and new model parameter estimates. α_i is defined as [187]

$$\alpha_i = \frac{n_i}{n_i + r} \tag{C.18}$$

where r is a fixed relevance factor, which determines the extent of mixing of the old and new estimates of the parameters. Low values for α_i ($\alpha_i \rightarrow 0$), will result in new parameter estimates from the data to be de-emphasized, while higher values ($\alpha_i \rightarrow 1$) will emphasize the use of the new training data-dependent parameters.

Generally only mean values are adapted [179]. It is experimentally shown that mean adaptation gives slightly higher performance than adapting all three parameters [187].

C.2 Testing

In identification phase, mixture densities are calculated for every feature vector for all speakers and speaker with maximum likelihood is selected as identified speaker. For example, if S speaker models $\{\Omega_1, \Omega_2, \ldots, \Omega_S\}$ are available after the training, speaker identification can be done based on a new speech data set. First, the sequence of feature vectors $X = \{x_1, x_2, \ldots, x_T\}$ is calculated. Then the speaker model $\hat{s}$ is determined which maximizes the a posteriori probability $P(\Omega_S|X)$. That is, according to the Bayes rule [178]

$$\hat{s} = \max_{1 \leq s \leq S} P(\Omega_S|X) = \max_{1 \leq s \leq S} \frac{P(X|\Omega_S)}{P(X)} P(\Omega_S). \qquad (C.19)$$

Assuming equal probability of all speakers and the statistical independence of the observations, the decision rule for the most probable speaker can be redefined as

$$\hat{s} = \max_{1 \leq s \leq S} \sum_{t=1}^{T} \log P(x_t|\Omega_s) \qquad (C.20)$$

with T the number of feature vectors of the speech data set under test and $P(x_t|\Omega_s)$ given by Eq. (C.1).

Decision in verification is obtained by comparing the score computed using the model for the claimed speaker Ω_S given by $P(\Omega_S|X)$ to a predefined threshold θ. The claim is accepted if $P(\Omega_S|X) > \theta$, and rejected otherwise [179].

References

1. D. Ververidis and C. Kotropoulos, "A state of the art review on emotional speech databases," in *Eleventh Australasian International Conference on Speech Science and Technology*, (Auckland, New Zealand), Dec. 2006.
2. S. G. Koolagudi, N. Kumar, and K. S. Rao, "Speech emotion recognition using segmental level prosodic analysis," in *International Conference on Devices and Communication*, (Mesra, India), Birla Institute of Technology, IEEE Press, Feb. 2011.
3. M.Schubiger, *English intonation: its form and function*. Tubingen, Germany: Niemeyer, 1958.
4. J. Connor and G.Arnold, *Intonation of Colloquial English*. London, UK: Longman, second ed., 1973.
5. M. E. Ayadi, M. S.Kamel, and F. Karray, "Survey on speech emotion recognition: Features,classification schemes, and databases," *Pattern Recognition*, vol. 44, pp. 572–587, 2011.
6. P. Ekman, *Handbook of Cognition and Emotion*, ch. Basic Emotions. Sussex, UK: John Wiley and Sons Ltd, 1999.
7. R.Cowie, E.Douglas-Cowie, N.Tsapatsoulis, S.Kollias, W.Fellenz, and J.Taylor, "Emotion recognition in human-computer interaction," *IEEE Signal Processing Magazine*, vol. 18, pp. 32–80, 2001.
8. J. William, "What is an emotion?," *Mind*, vol. 9, p. 188–205, 1984.
9. A. D. Craig, *Handbook of Emotion*, ch. Interoception and emotion: A neuroanatomical perspective. New York: The Guildford Press, September 2009. ISBN 978-1-59385-650-2.
10. C. E. Williams and K. N. Stevens, "Vocal correlates of emotional states," *Speech Evaluation in Psychiatry*, p. 189–220., 1981. Grune and Stratton Inc.
11. J.Cahn, "The generation of affect in synthesized speech," *Journal of American Voice Input/Output Society*, vol. 8, pp. 1–19, 1990.
12. G. M. David, "Theories of emotion," *Psychology*, vol. 7, 2004. New York, worth publishers.
13. X. Jin and Z. Wang, "An emotion space model for recognition of emotions in spoken chinese," in *ACII* (J. Tao, T. Tan, and R. Picard, eds.), pp. 397–402, LNCS 3784, Springer-Verlag Berlin Heidelberg, 2005.
14. J. Makhoul, "Linear prediction: A tutorial review," *Proceedings of the IEEE*, vol. 63, no. 4, pp. 561–580, 1975.
15. L. R. Rabiner and B. H. Juang, *Fundamentals of Speech Recognition*. Englewood Cliffs, New Jersy: Prentice-Hall, 1993.
16. J. Benesty, M. M. Sondhi, and Y. Huang, eds., *Springer Handbook on Speech Processing*. Springer Publishers, 2008.
17. S. G. Koolagudi and K. S. Rao, "Emotion recognition from speech using source, system and prosodic features," *International Journal of Speech Technology, Springer*, vol. 15, no. 3, pp. 265–289, 2012.

S.R. Krothapalli and S.G. Koolagudi, *Emotion Recognition using Speech Features*,   115
SpringerBriefs in Electrical and Computer Engineering, DOI 10.1007/978-1-4614-5143-3,
© Springer Science+Business Media New York 2013

18. M. Schroder, R. Cowie, E. Douglas-Cowie, M. Westerdijk, and S. Gielen, "Acoustic correlates of emotion dimensions in view of speech synthesis," (Aalborg, Denmark), EUROSPEECH 2001 Scandinavia, 2nd INTERSPEECH Event, September 3–7 2001. 7th European Conference on Speech Communication and Technology.

19. C.Williams and K.Stevens, "Emotionsandspeech:someacousticalcorrelates," *Journal of Acoustic Society of America*, vol. 52, no. 4 pt 2, pp. 1238–1250, 1972.

20. A. Batliner, J. Buckow, H. Niemann, E. Nöth, and VolkerWarnke, *Verbmobile Foundations of speech to speech translation.* ISBN 3540677836, 9783540677833: springer, 2000.

21. D. Ververidis and C. Kotropoulos, "Emotional speech recognition: Resources, features, and methods," *SPC*, vol. 48, p. 1162–1181, 2006.

22. F. Burkhardt, A. Paeschke, M. Rolfes, W. Sendlmeier, and B. Weiss, "A database of german emotional speech," in *Interspeech*, 2005.

23. S. G. Koolagudi, S. Maity, V. A. Kumar, S. Chakrabarti, and K. S. Rao, *IITKGP-SESC : Speech Database for Emotion Analysis.* Communications in Computer and Information Science, JIIT University, Noida, India: Springer, issn: 1865-0929 ed., August 17–19 2009.

24. E. McMahon, R. Cowie, S. Kasderidis, J. Taylor, and S. Kollias, "What chance that a dc could recognize hazardous mental states from sensor inputs?," in *Tales of the disappearing computer*, (Santorini , Greece), 2003.

25. C. M. Lee and S. S. Narayanan, "Toward detecting emotions in spoken dialogs," *IEEE Trans. Speech and Audio Processing*, vol. 13, pp. 293–303, March 2005.

26. B. Schuller, G. Rigoll, and M. Lang, "Speech emotion recognition combining acoustic features and linguistic information in a hybrid support vector machine-belief network architecture," in *IEEE International Conference on Acoustics, Speech, and Signal Processing, 2004. Proceedings. (ICASSP '04)*, (ISBN: 0-7803-8484-9), pp. I– 577–80, IEEE Press, May 17–21 2004.

27. F. Dellert, T. Polzin, and A. Waibel, "Recognizing emotion in speech," (Philadelphia, PA, USA), pp. 1970–1973, 4th International Conference on Spoken Language Processing, October 3–6 1996.

28. R. Nakatsu, J. Nicholson, and N. Tosa, "Emotion recognition and its application to computer agents with spontaneous interactive capabilities," *Knowledge-Based Systems*, vol. 13, pp. 497–504, December 2000.

29. F. Charles, D. Pizzi, M. Cavazza, T. Vogt, and E. André, "Emoemma: Emotional speech input for interactive storytelling," in *8th Int. Conf. on Autonomous Agents and Multiagent Systems (AAMAS 2009)* (Decker, Sichman, Sierra, and Castelfranchi, eds.), (Budapest, Hungary), pp. 1381–1382, International Foundation for Autonomous Agents and Multi-agent Systems, May, 10–15 2009.

30. T.V.Sagar, "Characterisation and synthesis of emotionsin speech using prosodic features," Master's thesis, Dept. of Electronics and communications Engineering, Indian Institute of Technology Guwahati, May. 2007.

31. D.J.France, R.G.Shiavi, S.Silverman, M.Silverman, and M.Wilkes, "Acoustical properties of speech as indicators of depression and suicidal risk," *IEEE Transactions on Biomedical Eng*, vol. 47, no. 7, pp. 829–837, 2000.

32. P.-Y. Oudeyer, "The production and recognition of emotions in speech: features and algorithms," *International Journal of Human Computer Studies*, vol. 59, p. 157–183, 2003.

33. J.Hansen and D.Cairns, "Icarus: source generator based real-time recognition of speech in noisy stressful and lombard effect environments," *Speech Communication*, vol. 16, no. 4, pp. 391–422, 1995.

34. M. Schroder and R. Cowie, "Issues in emotion-oriented computing – toward a shared understanding," in *Workshop on Emotion and Computing*, 2006. HUMAINE.

35. S. G. Koolagudi and K. S. Rao, "Real life emotion classification using vop and pitch based spectral features," in *INDICON-2010*, (KOLKATA-700032, INDIA), Jadavpur University, December 2010.

36. H. Wakita, "Residual energy of linear prediction to vowel and speaker recognition," *IEEE Trans. Acoust. Speech Signal Process*, vol. 24, pp. 270–271, 1976.

37. K. S. Rao, S. R. M. Prasanna, and B. Yegnanarayana, "Determination of instants of significant excitation in speech using hilbert envelope and group delay function," *IEEE Signal Processing Letters*, vol. 14, pp. 762–765, October 2007.

38. A. Bajpai and B. Yegnanarayana, "Exploring features for audio clip classification using lp residual and aann models," (Chennai, India), pp. 305–310, The international Conference on Intelligent Sensing and Information Processing 2004 (ICISIP 2004), January, 4–7 2004.

39. B. Yegnanarayana, R. K. Swamy, and K.S.R.Murty, "Determining mixing parameters from multispeaker data using speech-specific information," *IEEE Trans. Audio, Speech, and Language Processing*, vol. 17, no. 6, pp. 1196–1207, 2009. ISSN 1558–7916.

40. G. Seshadri and B. Yegnanarayana, "Perceived loudness of speech based on the characteristics of glottal excitation source," *Journal of Acoustic Society of America*, vol. 126, p. 2061–2071, October 2009.

41. K. E. Cummings and M. A. Clements, "Analysis of the glottal excitation of emotionally styled and stressed speech," *Journal of Acoustic Society of America*, vol. 98, pp. 88–98, July 1995.

42. L. Z. Hua and H. Y. andf Wang Ren Hua, "A novel source analysis method by matching spectral characters of lf model with straight spectrum." Springer-Verlag, Berlin, Heidelberg, 2005. 441–448.

43. D. O'Shaughnessy, *Speech Communication Human and Mechine*. Addison-Wesley publishing company, 1987.

44. M. Schröder, "Emotional speech synthesis: A review," in *7th European Conference on Speech Communication and Technology*, (Aalborg, Denmark), EUROSPEECH 2001 Scandinavia, September 3–7 2001.

45. S. G. Koolagudi and K. S. Rao, "Emotion recognition from speech : A review," *International Journal of Speech Technology, Springer*.

46. E. Douglas-Cowie, N. Campbell, R. Cowie, and P. Roach, "Emotional speech: Towards a new generation of databases," *SPC*, vol. 40, p. 33–60, 2003.

47. The 15th Oriental COCOSDA Conference, December 9–12, 2012, Macau, China. (http://www.ococosda2012.org/)

48. D. C. Ambrus, "Collecting and recording of an emotional speech database," tech. rep., Faculty of Electrical Engineering, Institute of Electronics, Univ. of Maribor, 2000.

49. M. Alpert, E. R. Pouget, and R. R. Silva, "Reflections of depression in acoustic measures of the patient's speech," *Journal of Affect Disord.*, vol. 66, pp. 59–69, September 2001.

50. A. Batliner, C. Hacker, S. Steidl, E. Noth, D. S. Archy, M. Russell, and M. Wong, "You stupid tin box – children interacting with the aibo robot: a cross-linguistic emotional speech corpus.," in *Proc. Language Resources and Evaluation (LREC '04)*, (Lisbon), 2004.

51. R. Cowie and E. Douglas-Cowie, "Automatic statistical analysis of the signal and prosodic signs of emotion in speech," in *Fourth International Conference on Spoken Language Processing (ICSLP '96)*,, (Philadelphia, PA, USA), pp. 1989–1992, October 1996.

52. R. Cowie and R. R. Cornelius, "Describing the emotional states that are expressed in speech," *Speech Communication*, vol. 40, pp. 5–32, Apr. 2003.

53. M. Edgington, "Investigating the limitations of concatenative synthesis," in *European Conference on Speech Communication and Technology (Eurospeech '97)*,, (Rhodes/Athens, Greece), pp. 593–596, 1997.

54. G. M. Gonzalez, "Bilingual computer-assisted psychological assessment: an innovative approach for screening depression in chicanos/latinos," tech. report-39, Univ. Michigan, 1999.

55. C. Pereira, "Dimensions of emotional meaning in speech," in *Proc. ISCA Workshop on Speech and Emotion*, (Belfast, Northern Ireland), pp. 25–28, 2000.

56. T. Polzin and A. Waibel, "Emotion sensitive human computer interfaces," in *ISCA Workshop on Speech and Emotion, Belfast*, pp. 201–206, 2000.

57. M. Rahurkar and J. H. L. Hansen, "Frequency band analysis for stress detection using a teager energy operator based feature," in *Proc. international conf. on spoken language processing(ICSLP'02)*, pp. Vol.3, 2021–2024, 2002.

58. K. R. Scherer, D. Grandjean, L. T. Johnstone, and T. B. G. Klasmeyer, "Acoustic correlates of task load and stress," in *International Conference on Spoken Language Processing (ICSLP '02),* (Colorado), pp. 2017–2020, 2002.

59. M. Slaney and G. McRoberts, "Babyears: A recognition system for affective vocalizations," *Speech Communication,* vol. 39, p. 367–384, February 2003.

60. S. Yildirim, M. Bulut, C. M. Lee, A. Kazemzadeh, C. Busso, Z. Deng., S. Lee, and S. Narayanan, "An acoustic study of emotions expressed in speech," (Jeju island, Korean), International Conference on Spoken Language Processing (ICSLP 2004), October 2004.

61. F. Burkhardt and W. F. Sendlmeier, "Verification of acousical correlates of emotional speech using formant-synthesis," (Newcastle, Northern Ireland, UK), pp. 151–156, ITRW on Speech and Emotion, September 5–7 2000.

62. A. Batliner, S. Biersacky, and S. Steidl, "The prosody of pet robot directed speech: Evidence from children," in *Speech Prosody 2006,* (Dresden), pp. 1–4, 2006.

63. M. Schroder and M. Grice, "Expressing vocal effort in concatenative synthesis," in *International Conference on Phonetic Sciences (ICPhS '03),* (Barcelona), 2003.

64. M. Schroder, "Experimental study of affect bursts," *Speech Communication - Special issue on speech and emotion,* vol. 40, no. 1–2, 2003.

65. M. Grimm, K. Kroschel, and S. Narayanan, "The vera am mittag german audio-visual emotional speech database," in *IEEE International Conference Multimedia and Expo,* (Hannover), pp. 865–868, April 2008. DOI: 10.1109/ICME.2008.4607572.

66. C. H. Wu, Z. J. Chuang, and Y. C. Lin, "Emotion recognition from text using semantic labels and separable mixture models," *ACM Transactions on Asian Language Information Processing (TALIP) TALIP,* vol. 5, pp. 165–182, June 2006.

67. T. L. Nwe, S. W. Foo, and L. C. D. Silva, "Speech emotion recognition using hidden Markov models," *Speech Communication,* vol. 41, pp. 603–623, Nov. 2003.

68. F. Yu, E. Chang, Y. Q. Xu, and H. Y. Shum, "Emotion detection from speech to enrich multimedia content," in *Proc. IEEE Pacific Rim Conference on Multimedia,* (Beijing), Vol.1 pp. 550–557, 2001.

69. J. Yuan, L. Shen, and F. Chen, "The acoustic realization of anger, fear, joy and sadness in chinese," in *International Conference on Spoken Language Processing (ICSLP '02),,* (Denver, Colorado, USA), pp. 2025–2028, September 2002.

70. I. Iriondo, R. Guaus, A. Rodríguez, P. Lázaro, N. Montoya, J. M. Blanco, D. Bernadas, J.M. Oliver, D. Tena, and L. Longhi, "Validation of an acoustical modeling of emotional expression in spanish using speech synthesis techniques," in *ITRW on Speech and Emotion,* (NewCastle, Northern Ireland, UK), September 2000. ISCA Archive.

71. J. M. Montro, J. Gutterrez-Arriola, J. Colas, E. Enriquez, and J. M. Pardo, "Analysis and modeling of emotional speech in spanish," in *Proc. Int.Conf. on Phonetic Sciences,* pp.957–960, 1999.

72. A. Iida, N. Campbell, F. Higuchi, and M. Yasumura, "A corpus-based speech synthesis system with emotion," *Speech Communication,* vol. 40, pp. 161–187, Apr. 2003.

73. V. Makarova and V. A. Petrushin, "Ruslana: A database of russian emotional utterances," in *International Conference on Spoken Language Processing (ICSLP '02),,* pp. 2041–2044, 2002.

74. M. Nordstrand, G. Svanfeldt, B. Granstrom, and D. House, "Measurements of ariculatory variation in expressive speech for a set of swedish vowels," *Speech Communication,* vol. 44, pp. 187–196, September 2004.

75. E. M. Caldognetto, P. Cosi, C. Drioli, G. Tisato, and F. Cavicchio, "Modifications of phonetic labial targets in emotive speech: effects of the co-production of speech and emotions," *Speech Communication,* vol. 44, no. 1–4, pp. 173–185, 2004.

76. J. Makhoul, "Linear prediction: A tutorial review," *Proc. IEEE,* vol. 63, pp. 561–580, Apr. 1975.

77. S. R. M. Kodukula, *Significance of Excitation Source Information for Speech Analysis.* PhD thesis, Dept. of Computer Science, IIT, Madras, March 2009.

78. T. V. Ananthapadmanabha and B. Yegnanarayana, "Epoch extraction from linear prediction residual for identification of closed glottis interval," *IEEE Trans. Acoustics, Speech, and Signal Processing*, vol. 27, pp. 309–319, Aug. 1979.

79. B.Yegnanarayana, S.R.M.Prasanna, and K. Rao, "Speech enhancement using excitation source information," in *Proc. IEEE Int. Conf. Acoust., Speech, Signal Processing*, vol. 1, (Orlando, Florida, USA), pp. 541–544, May 2002.

80. A. Bajpai and B.Yegnanarayana, "Combining evidence from sub-segmental and segmental features for audio clip classification," in *IEEE Region 10 Conference (TENCON)*, (India), pp. 1–5, IIIT, Hyderabad, Nov. 2008.

81. B. S. Atal, "Automatic speaker recognition based on pitch contours," *Journal of Acoustic Society of America*, vol. 52, no. 6, pp. 1687–1697, 1972.

82. P. Thevenaz and H. Hugli, "Usefulness of lpc residue in textindependent speaker verification," *Speech Communication*, vol. 17, pp. 145–157, 1995.

83. J. H. L. Liu and G. Palm, "On the use of features from prediction residual signal in speaker recognition," pp. 313–316, Proc. European Conf. Speech Processing, Technology (EUROSPEECH), 1997.

84. B. Yegnanarayana, P. S. Murthy, C. Avendano, and H. Hermansky, "Enhancement of reverberant speech using lp residual," in *IEEE International Conference on Acoustics, Speech and Signal Processing*, (Seattle, WA , USA), pp. 405–408 vol.1, IEEE Xplore, May 1998. DOI:10.1109/ICASSP.1998.674453.

85. K. S. Kumar, M. S. H. Reddy, K. S. R. Murty, and B. Yegnanarayana, "Analysis of laugh signals for detecting in continuous speech," (Brighton, UK), pp. 1591–1594, INTERSPEECH, September, 6–10 2009.

86. G. Bapineedu, B. Avinash, S. V. Gangashetty, and B. Yegnanarayana, "Analysis of lombard speech using excitation source information," (Brighton, UK), pp. 1091–1094, INTER-SPEECH, September, 6–10 2009.

87. O. M. Mubarak, E. Ambikairajah, and J. Epps, "Analysis of an mfcc-based audio indexing system for efficient coding of multimedia sources," in *The 8th International Symposium on Signal Processing and its Applications*, (Sydney, Australia), 28–31 August 2005.

88. T. L. Pao, Y. T. Chen, J. H. Yeh, and W. Y. Liao, "Combining acoustic features for improved emotion recognition in mandarin speech," in *ACII* (J. Tao, T. Tan, and R. Picard, eds.), (LNCS 3784), pp. 279–285, ⓒSpringer-Verlag Berlin Heidelberg, 2005.

89. T. L. Pao, Y. T. Chen, J. H. Yeh, Y. M. Cheng, and C. S. Chien, *Feature Combination for Better Differentiating Anger from Neutral in Mandarin Emotional Speech*. LNCS 4738, ACII 2007: Springer-Verlag Berlin Heidelberg, 2007.

90. N. Kamaruddin and A. Wahab, "Features extraction for speech emotion," *Journal of Computational Methods in Science and Engineering*, vol. 9, no. 9, pp. 1–12, 2009. ISSN:1472–7978 (Print) 1875–8983 (Online).

91. D. Neiberg, K. Elenius, and K. Laskowski, "Emotion recognition in spontaneous speech using GMMs," in *INTERSPEECH 2006 - ICSLP*, (Pittsburgh, Pennsylvania), pp. 809–812, 17–19 September 2006.

92. D. Bitouk, R. Verma, and A. Nenkova, "Class-level spectral features for emotion recognition," *Speech Communication*, 2010. Article in press.

93. M. Sigmund, "Spectral analysis of speech under stress," *IJCSNS International Journal of Computer Science and Network Security*, vol. 7, pp. 170–172, April 2007.

94. K. S. Rao and B. Yegnanarayana, "Prosody modification using instants of significant excitation," *IEEE Trans. Speech and Audio Processing*, vol. 14, pp. 972–980, May 2006.

95. S. Werner and E. Keller, "Prosodic aspects of speech," in *Fundamentals of Speech Synthesis and Speech Recognition: Basic Concepts, State of the Art, the Future Challenges* (E. Keller, ed.), pp. 23–40, Chichester: John Wiley, 1994.

96. T. Banziger and K. R. Scherer, "The role of intonation in emotional expressions," *Speech Communication*, no. 46, pp. 252–267, 2005.

97. R. Cowie and R. R. Cornelius, "Describing the emotional states that are expressed in speech," *Speech Communication*, vol. 40, pp. 5–32, Apr. 2003.

98. F. Dellaert, T. Polzin, and A. Waibel, "Recognising emotions in speech," ICSLP 96, Oct. 1996.

99. M. Schroder, "Emoptional speech synthesis: A review," (Seventh european conference on speech communication and technology Aalborg, Denmark), Eurospeech 2001, Sept. 2001.

100. I. R. Murray and J. L. Arnott, "Implementation and testing of a system for producing emotion by rule in synthetic speech," *Speech Communication*, vol. 16, pp. 369–390, 1995.

101. J. E. Cahn, "The generation of affect in synthesized speech," *JAVIOS*, pp. 1–19, Jul. 1990.

102. I. R. Murray, J. L. Arnott, and E. A. Rohwer, "Emotional stress in synthetic speech: Progress and future directions," *Speech Communication*, vol. 20, pp. 85–91, Nov. 1996.

103. K. R. Scherer, "Vocal communication of emotion: A review of research paradigms," *Speech Communication*, vol. 40, pp. 227–256, 2003.

104. S. McGilloway, R. Cowie, E. Douglas-Cowie, S. Gielen, M. Westerdijk, and S. Stroeve, "Approaching automatic recognition of emotion from voice: A rough benchmark," (Belfast), 2000.

105. I. Luengo, E. Navas, I. Hernáez, and J. Sánchez, "Automatic emotion recognition using prosodic parameters," in *INTERSPEECH*, (Lisbon, Portugal), pp. 493–496, IEEE, September 2005.

106. T. Iliou and C.-N. Anagnostopoulos, "Statistical evaluation of speech features for emotion recognition," in *Fourth International Conference on Digital Telecommunications*, (Colmar, France), pp. 121–126, July 2009. ISBN: 978-0-7695-3695-8.

107. Y. hao Kao and L. shan Lee, "Feature analysis for emotion recognition from mandarin speech considering the special characteristics of chinese language," in *INTERSPEECH -ICSLP*, (Pittsburgh, Pennsylvania), pp. 1814–1817, September 2006.

108. A. Zhu and Q. Luo, "Study on speech emotion recognition system in e learning," in *Human Computer Interaction, Part III, HCII* (J. Jacko, ed.), (Berlin Heidelberg), pp. 544–552, Springer Verlag, 2007. LNCS:4552, DOI: 10.1007/978-3-540-73110-8-59.

109. M. Lugger and B. Yang, "The relevance of voice quality features in speaker independent emotion recognition," in *ICASSP*, (Honolulu, Hawai, USA), pp. IV17–IV20, IEEE, May 2007.

110. Y. Wang, S. Du, and Y. Zhan, "Adaptive and optimal classification of speech emotion recognition," in *Fourth International Conference on Natural Computation*, pp. 407–411, October 2008. http://doi.ieeecomputersociety.org/10.1109/ICNC.2008.713.

111. S. Zhang, "Emotion recognition in chinese natural speech by combining prosody and voice quality features," in *Advances in Neural Networks, Lecture Notes in Computer Science, Volume 5264* (S. et al., ed.), (Berlin Heidelberg), pp. 457–464, Springer Verlag, 2008. DOI: 10.1007/978-3-540-87734-9-52.

112. D. Ververidis, C. Kotropoulos, and I. Pitas, "Automatic emotional speech classification," pp. I593–I596, ICASSP 2004, IEEE, 2004.

113. K. S. Rao, R. Reddy, S. Maity, and S. G. Koolagudi, "Characterization of emotions using the dynamics of prosodic features," in *International Conference on Speech Prosody*, (Chicago, USA), May 2010.

114. K. S. Rao, S. R. M. Prasanna, and T. V. Sagar, "Emotion recognition using multilevel prosodic information," in *Workshop on Image and Signal Processing (WISP-2007)*, (Guwahati, India), IIT Guwahati, Guwahati, December 2007.

115. Y.Wang and L.Guan, "An investigation of speech-based human emotion recognition," in *IEEE 6th Workshop on Multimedia Signal Processing*, pp. 15–18, IEEE press, October 2004.

116. Y. Zhou, Y. Sun, J. Zhang, and Y. Yan, "Speech emotion recognition using both spectral and prosodic features," in *International Conference on Information Engineering and Computer Science, ICIECS*, (Wuhan), pp. 1–4, IEEE press, 19–20 Dec. 2009. DOI: 10.1109/ICIECS.2009.5362730.

117. C. E. X. Y. Yu, F. and H. Shum, "Emotion detection from speech to enrich multimedia content," in *Second IEEE Pacific-Rim Conference on Multimedia*, (Beijing, China), October 2001.

118. V.Petrushin, *Emotion in speech: Recognition and application to call centres.* Artifi.Neu.Net. Engr.(ANNIE), 1999.

119. R. Nakatsu, J. Nicholson, and N. Tosa, "Emotion recognition and its application to computer agents with spontaneous interactive capabilities," *Knowledge Based Systems*, vol. 13, pp.497–504, 2000.

120. J. Nicholson, K. Takahashi, and R.Nakatsu, "Emotion recognition in speech using neural networks," *Neural computing and applications*, vol. 11, pp. 290–296, 2000.

121. R. Tato, R. Santos, R. Kompel, and J. Pardo, "Emotional space improves emotion recognition," (Denver, Colorado, USA), 7th International Conference on Spoken Language Processing, September 16–20 2002.

122. R. Fernandez and R. W. Picard, "Modeling drivers' speech under stress," *Speech Communication*, vol. 40, p. 145–159, 2003.

123. V. A. Petrushin, "Emotion in speech : Recognition and application to call centers," Proceedings of the 1999 Conference on Artificial Neural Networks in Engineering (ANNIE '99), 1999.

124. J. Nicholson, K. Takahashi, and R.Nakatsu, "Emotion recognition in speech using neural networks," in *6th International Conference on Neural Information Processing*, (Perth, WA, Australia), pp. 495–501, ICONIP-99, August 1999. 10.1109/ICONIP.1999.845644.

125. V. A. Petrushin, "Emotion recognition in speech signal: Experimental study, development and application," in *ICSLP*, (Beijing, China), 2000.

126. C. M. Lee, S. Narayanan, and R. Pieraccini, "Recognition of negative emotion in the human speech signals," in *Workshop on Auto. Speech Recognition and Understanding*, December 2001.

127. G. Zhou, J. H. L. Hansen, and J. F. Kaiser, "Nonlinear feature based classification of speech under stress," *IEEE Trans. Speech and Audio Processing*, vol. 9, pp. 201–216, March 2001.

128. K. S. Rao and S. G. Koolagudi, "Characterization and recognition of emotions from speech using excitation source information," *International Journal of Speech Technology, Springer.* DOI 10.1007/s10772-012-9175-2.

129. K. S. R. Murty and B. Yegnanarayana, "Combining evidence from residual phase and mfcc features for speaker recognition," *IEEE SIGNAL PROCESSING LETTERS*, vol. 13, pp.52–55, January 2006.

130. K. Murty and B. Yegnanarayana, "Epoch extraction from speech signals," *IEEE Trans. Audio, Speech, and Language Processing*, vol. 16, pp. 1602–1613, 2008.

131. B. Yegnanarayana, *Artificial Neural Networks.* New Delhi, India: Prentice-Hall, 1999.

132. S. Haykin, *Neural Networks: A Comprehensive Foundation.* New Delhi, India: Pearson Education Aisa, Inc., 1999.

133. K. S. Rao, "Role of neural network models for developing speech systems," *Sadhana, Academy Proceedings in Engineering Sciences, Indian Academy of Sciences, Springer*, vol. 36, pp. 783–836, Oct. 2011.

134. R. H. Laskar, D. Chakrabarty, F. A. Talukdar, K. S. Rao, and K. Banerjee, "Comparing ANN and GMM in a voice conversion framework," *Applied Soft Computing,Elsevier*, vol. 12, pp. 3332–3342, Nov. 2012.

135. K. I. Diamantaras and S. Y. Kung, *Principal Component Neural Networks: Theory and Applications.* Newyork: John Wiley and Sons, 1996.

136. M. S. Ikbal, H. Misra, and B. Yegnanarayana, "Analysis of autoassociative mapping neural networks," (USA), pp. 854–858, Proc. Internat. Joint Conf. on Neural Networks (IJCNN), 1999.

137. S. P. Kishore and B. Yegnanarayana, "Online text-independent speaker verification system using autoassociative neural network models," (Washington, DC, USA.), pp. 1548–1553 (V2), Proc. Internat. Joint Conf. on Neural Networks (IJCNN), August 2001.

138. A. V. N. S. Anjani, "Autoassociate neural network models for processing degraded speech," Master's thesis, MS thesis, Department of Computer Science and Engineering, Indian Institute of Technology Madras, Chennai 600 036, India, 2000.

139. K. S. Reddy, "Source and system features for speaker recognition," Master's thesis, MS thesis, Department of Computer Science and Engineering, Indian Institute of Technology Madras, Chennai 600 036, India, 2004.

140. C. S. Gupta, "Significance of source features for speaker recognition," Master's thesis, MS thesis, Department of Computer Science and Engineering, Indian Institute of Technology Madras, Chennai 600 036, India, 2003.

141. S. Desai, A. W. Black, B.Yegnarayana, and K. Prahallad, "Spectral mapping using artificial neural networks for voice conversion," *IEEE Trans. Audio, Speech, and Language Processing*, vol. 18, pp. 954–964, 8 Apr. 2010.

142. K. S. Rao and B. Yegnanarayana, "Intonation modeling for indian languages," *Computer Speech and Language*, vol. 23, pp. 240–256, April 2009.

143. C. K. Mohan and B. Yegnanarayana, "Classification of sport videos using edge-based features and autoassociative neural network models," *Signal, Image and Video Processing*, vol. 4, pp. 61–73, 15 Nov. 2008. DOI: 10.1007/s11760-008-0097-9.

144. L. Mary and B. Yegnanarayana, "Autoassociative neural network models for language identification," in *International Conference on Intelligent Sensing and Information Processing*, pp. 317–320, IEEE, 24 Aug. 2004. DOI:10.1109/ICISIP.2004.1287674.

145. K. S. Rao, J. Yadav, S. Sarkar, S. G. Koolagudi, and A. K. Vuppala, "Neural network based feature transformation for emotion independent speaker identification," *International Journal of Speech Technology, Springer*, vol. 15, no. 3, pp. 335–349, 2012.

146. B. Yegnanarayana, K. S. Reddy, and S. P. Kishore, "Source and system features for speaker recognition using aann models," (Salt Lake City, UT), IEEE Int. Conf. Acoust., Speech, and Signal Processing, May 2001.

147. C. S. Gupta, S. R. M. Prasanna, and B. Yegnanarayana, "Autoassociative neural network models for online speaker verification using source features from vowels," in *Int. Joint Conf. Neural Networks*, (Honululu, Hawii, USA), May 2002.

148. B. Yegnanarayana, K. S. Reddy, and S. P. Kishore, "Source and system features for speaker recognition using AANN models," in *Proc. IEEE Int. Conf. Acoust., Speech, Signal Processing*, (Salt Lake City, Utah, USA), pp. 409–412, May 2001.

149. S. Theodoridis and K. Koutroumbas, *Pattern Recognition*. USA: Elsevier, Academic press, 3 ed., 2006.

150. K. S. Rao, *Acquisition and incorporation prosody knowledge for speech systems in Indian languages*. PhD thesis, Dept. of Computer Science and Engineering, Indian Institute of Technology Madras, Chennai, India, May 2005.

151. S. R. M. Prasanna, B. V. S. Reddy, and P. Krishnamoorthy, "Vowel onset point detection using source, spectral peaks, and modulation spectrum energies," *IEEE Trans. Audio, Speech, and Language Processing*, vol. 17, pp. 556–565, March 2009.

152. S. G. Koolagudi and K. S. Rao, "Emotion recognition from speech using sub-syllabic and pitch synchronous spectral features," *International Journal of Speech Technology, Springer*. DOI 10.1007/s10772-012-9150-8.

153. J. Chen, Y. A. Huang, Q. Li, and K. K. Paliwal, "Recognition of noisy speech using dynamic spectral subband centroids," *IEEE signal processing letters*, vol. 11, pp. 258–261, February 2004.

154. B. Yegnanarayana and S. P. Kishore, "AANN an alternative to GMM for pattern recognition," *Neural Networks*, vol. 15, pp. 459–469, Apr. 2002.

155. R. O. Duda, P. E. Hart, and D. G. Stork, *Pattern Classification*. Singapore: A Wiley-interscience Publications, 2 ed., 2004.

156. S. R. M. Prasanna, B. V. S. Reddy, and P. Krishnamoorthy, "Vowel onset point detection using source, spectral peaks, and modulation spectrum energies," *IEEE Trans. Audio, Speech, and Language Processing*, vol. 17, pp. 556–565, May 2009.

157. Unicode Entity Codes for the Telugu Script, Accents, Symbols and Foreign Scripts, Penn State University, USA. (http://tlt.its.psu.edu/suggestions/international/bylanguage/teluguchart.html)

158. K. S. Rao, *Predicting Prosody from Text for Text-to-Speech Synthesis*. ISBN-13: 978-1461413370, Springer, 2012.

159. K. S. Rao and S. G. Koolagudi, "Selection of suitable features for modeling the durations of syllables," *Journal of Software Engineering and Applications*, vol. 3, pp. 1107–1117, Dec. 2010.

160. K. S. Rao, "Application of prosody models for developing speech systems in indian languages," *International Journal of Speech Technology, Springer*, vol. 14, pp. 19–33, 2011.

161. N. P. Narendra, K. S. Rao, K. Ghosh, R. R. Vempada, and S. Maity, "Development of syllable-based text-to-speech synthesis system in bengali," *International Journal of Speech Technology, Springer*, vol. 14, no. 3, pp. 167–181, 2011.

162. K. S. Rao, S. G. Koolagudi, and R. R. Vempada, "Emotion recognition from speech using global and local prosodic features," *International Journal of Speech Technology, Springer*, Aug. 2012. DOI: 10.1007/s10772-012-9172-2.

163. L. R. Rabiner, *Digital Signal Processing*. IEEE Press, 1972.

164. B. S. Atal and S. L. Hanauer, "Speech analysis and synthesis by linear prediction of the speech wave," *J. Acoust. Soc. Am.*, vol. 50, pp. 637–655, Aug. 1971.

165. J. Makhoul, "Linear prediction: A tutorial review," *Proc. IEEE*, vol. 63, pp. 561–580, Apr. 1975.

166. B. S. Atal and M. R. Schroeder, "Linear prediction analysis of speech based on a pole-zero representation," *J. Acoust. Soc. Am.*, vol. 64, no. 5, pp. 1310–1318, 1978.

167. D. O'Shaughnessy, "Linear predictive coding," *IEEE Potentials*, vol. 7, pp. 29–32, Feb. 1988.

168. T. Ananthapadmanabha and B. Yegnanarayana, "Epoch extraction from linear prediction residual for identification of closed glottis interval," *IEEE Trans. Acoust., Speech, Signal Process.*, vol. ASSP-27, pp. 309–319, Aug. 1979.

169. J. Picone, "Signal modeling techniques in speech recognition," *Proc. IEEE*, vol. 81, pp.1215–1247, Sep. 1993.

170. J. W. Picone, "Signal modeling techniques in speech recognition," *Proceedings of IEEE*, vol. 81, pp. 1215–1247, Sep. 1993.

171. J. R. Deller, J. H. Hansen, and J. G. Proakis, *Discrete Time Processing of Speech Signals*. 1st ed. Upper Saddle River, NJ, USA: Prentice Hall PTR, 1993.

172. J. Benesty, M. M. Sondhi, and Y. A. Huang, *Springer Handbook of Speech Processing*. Springer-Verlag New York, Inc., 2008.

173. J. Volkmann, S. Stevens, and E. Newman, "A scale for the measurement of the psychological magnitude pitch," *J. Acoust. Soc. Amer.*, vol. 8, pp. 185–190, Jan. 1937.

174. Z. Fang, Z. Guoliang, and S. Zhanjiang, "Comparison of different implementations of MFCC," *J. Computer Science and Technology*, vol. 16, no. 6, pp. 582–589, 2001.

175. G. K. T. Ganchev and N. Fakotakis, "Comparative evaluation of various MFCC implementations on the speaker verification task," in *Proc. of Int. Conf. on Speech and Computer*, (Patras, Greece), pp. 191–194, 2005.

176. S. Furui, "Comparison of speaker recognition methods using statistical features and dynamic features," *IEEE Trans. Acoust., Speech, Signal Process.*, vol. 29, no. 3, pp. 342–350, 1981.

177. J. S. Mason and X. Zhang, "Velocity and acceleration features in speaker recognition," in *Proc. IEEE Int. Conf. Acoust., Speech, Signal Processing*, (Toronto, Canada), pp. 3673–3676, Apr. 1991.

178. D. A. Reynolds, "Speaker identification and verification using Gaussian mixture speaker models," *Speech Communication*, vol. 17, pp. 91–108, Aug. 1995.

179. F. Bimbot, J. F. Bonastre, C. Fredouille, G. Gravier, M. I. Chagnolleau, S. Meignier, T. Merlin, O. J. Garcia, D. Petrovska, and Reynolds, "A tutorial on text-independent speaker verification," *EURASIP Journal Applied Signal process*, no. 4, pp. 430–451, 2004.

180. A. Dempster, N. M. Laird, and D. B. Rubin, "Maximum likelihood from incomplete data via the EM algorithm," *Journal Royal Statistical Society*, vol. 39, no. 1, pp. 1–38, 1977.

181. Y. Linde, A. Buzo, and R. Gray, "An algorithm for vector quantizer design," *IEEE Trans. Communications*, vol. 28, pp. 84–95, Jan. 1980.

182. J. B. MacQueen, "Some methods for classification and analysis of multivariate observations," in *Proc. of the fifth Berkeley Symposium on Mathematical Statistics and Probability* (L. M. L. Cam and J. Neyman, eds.), vol. 1, pp. 281–297, University of California Press, 1967.

183. J. A. Hartigan and M. A. Wong, "A K-means clustering algorithm," *Applied Statistics*, vol. 28, no. 1, pp. 100–108, 1979.

184. Q. Y. Hong and S. Kwong, "A discriminative training approach for text-independent speaker recognition," *Signal process.*, vol. 85, no. 7, pp. 1449–1463, 2005.

185. D. Reynolds and R. Rose, "Robust text-independent speaker identification using Gaussian mixture speaker models," *IEEE Trans. Speech Audio processeing*, vol. 3, pp. 72–83, Jan. 1995.

186. J. Gauvain and C.-H. Lee, "Maximum a posteriori estimation for multivariate Gaussian mixture observations of Markov chains," *IEEE Trans. Speech Audio process.*, vol. 2, pp.291–298, Apr. 1994.

187. D. A. Reynolds, "Speaker verification using adapted Gaussian mixture models," *Digital Signal Process.*, vol. 10, pp. 19–41, Jan. 2000.